A PRIMER ON STATISTICAL DISTRIBUTIONS

A PRIMER ON STATISTICAL DISTRIBUTIONS

N. BALAKRISHNAN
McMaster University
Hamilton, Canada

V. B. NEVZOROV
St. Petersburg State University
Russia

WILEY-
INTERSCIENCE

A JOHN WILEY & SONS, INC., PUBLICATION

Library of Congress Cataloging-in-Publication Data:

Balakrishnan, N., 1956–
    A primer on statistical distributions / N. Balakrishnan and V.B. Nevzorov.
        p. cm.
    Includes bibliographical references and index.
    ISBN 0-471-42798-5 (acid-free paper)
    1. Distribution (Probability theory)   I. Nevzorov, Valery B., 1946–   II. Title.

    QA273.B25473 2003
    519.2'4—dc21   2003041157

Printed in the United States of America.

10 9 8 7 6 5 4 3 2 1

To my lovely daughters, Sarah and Julia
(N.B.)

To my wife, Ludmila
(V.B.N.)

CONTENTS

III MULTIVARIATE DISTRIBUTIONS 247

PREFACE

Distributions and their properties and interrelationships assume a very important role in most upper-level undergraduate as well as graduate courses in the statistics program. For this reason, many introductory statistics textbooks discuss in a chapter or two a few basic statistical distributions, such as binomial, Poisson, exponential, and normal. Yet a good knowledge of some other distributions, such as geometric, negative binomial, Pareto, beta, gamma, chi-square, logistic, Laplace, extreme value, multinomial, multivariate normal, and Dirichlet will be immensely useful to those students who go on to upper-level undergraduate or graduate courses in statistics. Students in applied programs such as psychology, sociology, biology, geography, geology, economics, business, and engineering will also benefit significantly from an exposure to different distributions and their properties as statistical modelling of observed data is an integral part of these disciplines.

It is for this reason we have prepared this textbook, which is tailor-made for a one-term course (of about 35 lectures) on *statistical distributions*. All the preliminary concepts and definitions are presented in Chapter 1. The rest of the material is divided into three parts, with Part I covering discrete distributions, Part II covering continuous distributions, and Part III covering multivariate distributions. In each chapter we have included a few pertinent exercises (at an appropriate level for students taking the course) which may be handed out as homework at the end of each chapter. A biographical sketch of some of the leading contributors to the area of *statistical distribution theory* is presented in the Appendix to present students with a historical sense of developments in this important and fundamental area in the field of statistics.

From our experience, we would suggest the following lecture allocation for teaching a course on *statistical distributions* based on this book:

5	**lectures**	on	*preliminaries*	(Chapter 1)
9	**lectures**	on	*discrete distributions*	(Part I)
17	**lectures**	on	*continuous distributions*	(Part II)
4	**lectures**	on	*multivariate distributions*	(Part III)

We welcome comments and criticisms from all those who teach a course based on this book. Any suggestions for improvement or "necessary" addition (omission of which in this version should be regarded as a consequence of our

xv

ignorance, not of personal nonscientific antipathy) sent to us will be much appreciated and will be acted upon when the opportunity arises.

It is important to mention here that many authoritative and encyclopedic volumes on statistical distribution theory exist in the literature. For example:

- Johnson, Kotz, and Kemp (1992), describing discrete univariate distributions

- Stuart and Ord (1993), discussing general distribution theory

- Johnson, Kotz, and Balakrishnan (1994, 1995), describing continuous univariate distributions

- Johnson, Kotz, and Balakrishnan (1997), describing discrete multivariate distributions

- Wimmer and Altmann (1999), providing a thesaurus on discrete univariate distributions

- Evans, Peacock, and Hastings (2000), describing discrete and continuous distributions

- Kotz, Balakrishnan, and Johnson (2000), discussing continuous multivariate distributions

are some of the prominent ones. In addition, there are separate books dedicated to some specific distributions, such as Poisson, generalized Poisson, chi-square, Pareto, exponential, lognormal, logistic, normal, and Laplace (which have all been referred to in this book at appropriate places). These books may be consulted for any additional information.

We take this opportunity to express our sincere thanks to Mr. Steve Quigley (of John Wiley & Sons, New York) for his support and encouragement during the preparation of this book. Our special thanks go to Mrs. Debbie Iscoe (Mississauga, Ontario, Canada) for assisting us with the camera-ready production of the manuscript, and to Mr. Weiquan Liu for preparing all the figures. We also acknowledge with gratitude the financial support provided by the Natural Sciences and Engineering Research Council of Canada and the Russian Foundation of Basic Research (Grants 01-01-00031 and 00-15-96019) during the course of this project.

N. BALAKRISHNAN
Hamilton, Canada

V. B. NEVZOROV
St. Petersburg, Russia

April 2003

CHAPTER 1

PRELIMINARIES

In this chapter we present some basic notations, notions, and definitions which a reader of this book must absolutely know in order to follow subsequent chapters.

1.1 Random Variables and Distributions

Let $(\Omega, \mathcal{T}, P)$ be a probability space, where $\Omega = \{\omega\}$ is a set of elementary events, $\mathcal{T}$ is a σ-algebra of events, and P is a probability measure defined on $(\Omega, \mathcal{T})$. Further, let B denote an element of the Borel σ-algebra of subsets of the real line $\mathcal{R}$.

Definition 1.1 A finite single-valued function $X = X(\omega)$ which maps Ω into $\mathcal{R}$ is called a *random variable* if for any Borel set B in $\mathcal{R}$, the inverse image of B, i.e.,

$$X^{-1}(B) = \{\omega : X(\omega) \in B\}$$

belongs to the σ-algebra $\mathcal{T}$.

It means that for all Borel sets B, one can define probabilities

$$P\{X \in B\} = P\{X^{-1}(B)\}.$$

In particular, if for any x $(-\infty < x < \infty)$ we take $B = (-\infty, x]$, then the function

$$F(x) = P\{X \le x\} \tag{1.1}$$

is defined for the random variable X.

Definition 1.2 The function $F(x)$ is called the *distribution function* or *cumulative distribution function* (cdf) of the random variable X.

Remark 1.1 Quite often, the cumulative distribution function of a random variable X is defined as

$$G(x) = P\{X < x\}.$$

1

Most of the properties of both these versions of cdf (i.e., F and G) coincide. Only one important difference exists between functions $F(x)$ and $G(x)$: F is right continuous, while G is left continuous. In our treatment we use the cdf as given in Definition 1.2.

There are three types of distributions: absolutely continuous, discrete and singular, and any cdf $F(x)$ can be represented as a mixture

$$F(x) = p_1 F_1(x) + p_2 F_2(x) + p_3 F_3(x) \qquad (1.2)$$

of absolutely continuous F_1, discrete F_2, and singular F_3 cdf's, with non-negative weights p_1, p_2, and p_3 such that $p_1 + p_2 + p_3 = 1$. In this book we restrict ourselves to distributions which are either purely absolutely continuous or purely discrete.

Definition 1.3 A random variable X is said to have a *discrete distribution* if there exists a countable set $B = \{x_1, x_2, \ldots\}$ such that

$$P\{X \in B\} = 1.$$

Remark 1.2 To determine a random variable having a discrete distribution, one must fix two sequences: a sequence of values $x_1, x_2, \ldots$ and a sequence of probabilities $p_k = P\{X = x_k\}$, $k = 1, 2, \ldots$, such that

$$\sum_k p_k = 1.$$

In this case, the cdf of X is given by

$$F(x) = P\{X \le x\} = \sum_{k:\ x_k \le x} p_k. \qquad (1.3)$$

Definition 1.4 A random variable X with a cdf F is said to have an *absolutely continuous distribution* if there exists a nonnegative function $p(x)$ such that

$$F(x) = \int_{-\infty}^{x} p(t)\ dt \qquad (1.4)$$

for any real x.

Remark 1.3 The function $p(x)$ then satisfies the condition

$$\int_{-\infty}^{\infty} p(t)\ dt = 1, \qquad (1.5)$$

and it is called the *probability density function* (pdf) of X. Note that any nonnegative function $p(x)$ satisfying (1.5) can be the pdf of some random variable X.

Remark 1.4 If a random variable X has an absolutely continuous distribution, then its cdf $F(x)$ is continuous.

Definition 1.5 We say that random variables X and Y have the same distribution, and write

$$X \overset{d}{=} Y \tag{1.6}$$

if the cdf's of X and Y (i.e., F_X and F_Y) coincide; that is,

$$F_X(x) = P\{X \leq x\} = P\{Y \leq x\} = F_Y(x) \ \forall \ x.$$

Exercise 1.1 Construct an example of a probability space $(\Omega, \mathcal{T}, P)$ and a finite single-valued function $X = X(\omega), \omega \in \Omega$, which maps Ω into $\mathcal{R}$, that is not a random variable.

Exercise 1.2 Let $p(x)$ and $q(x)$ be probability density functions of two random variables. Consider now the following functions:

(a) $2p(x) - q(x)$; (b) $p(x) + 2q(x)$; (c) $|p(x) - q(x)|$; (d) $\dfrac{1}{2}(p(x) + q(x))$.

Which of these functions are probability density functions of some random variable for any choice of $p(x)$ and $q(x)$? Which of them can br valid probability density functions under suitably chosen $p(x)$ and $q(x)$? Is there a function that can never be a probability density function of a random variable?

Exercise 1.3 Suppose that $p(x)$ and $q(x)$ are probability density functions of X and Y, respectively, satisfying

$$p(x) = 2 - q(x) \quad \text{for} \quad 0 < x < 1.$$

Then, find $P\{X < -1\} + P\{Y < 2\}$.

The *quantile function* of a random variable X with cdf $F(x)$ is defined by

$$Q(u) = \inf\{x : F(x) \geq u\}, \quad 0 < u < 1.$$

In the case when X has an absolutely continuous distribution, then the quantile function $Q(u)$ may simply be written as

$$Q(u) = F^{-1}(u), \quad 0 < u < 1.$$

The corresponding *quantile density function* is given by

$$q(u) = \frac{dQ(u)}{du} = \frac{1}{p(Q(u))}, \quad 0 < u < 1,$$

where $p(x)$ is the pdf corresponding to the cdf $F(x)$.

It should be noted that just as forms of $F(x)$ may be used to propose families of distributions, general forms of the quantile function $Q(u)$ may also be used to propose families of statistical distributions. Interested readers may refer to the recent book by Gilchrist (2000) for a detailed discussion on statistical modelling with quantile functions.

1.2   Type of Distribution

Definition 1.6 Random variables X and Y are said to *belong to the same type of distribution* if there exist constants a and $h > 0$ such that

$$Y \stackrel{d}{=} a + hX. \tag{1.7}$$

Note then that the cdf's F_X and F_Y of the random variables X and Y satisfy the relation

$$F_Y(x) = F_X\left(\frac{x - a}{h}\right) \ \forall \ x. \tag{1.8}$$

One can, therefore, choose a certain cdf F as the standard distribution function of a certain distribution family. Then this family would consist of all cdf's of the form

$$F(x, a, h) = F\left(\frac{x - a}{h}\right), \qquad -\infty < x < \infty, \ h > 0, \tag{1.9}$$

and

$$F(x) = F(x, 0, 1).$$

Thus, we have a two-parameter family of cdf's $F(x, a, h)$, where a is called the *location parameter* and h is the *scale parameter*.

For absolutely continuous distributions, one can introduce the corresponding two-parameter families of probability density functions:

$$p(x, a, h) = \frac{1}{h} \, p\left(\frac{x - a}{h}\right), \tag{1.10}$$

where $p(x) = p(x, 0, 1)$ corresponds to the random variable X with cdf F, and $p(x, a, h)$ corresponds to the random variable $Y = a + hX$ with cdf $F(x, a, h)$.

1.3   Moment Characteristics

There are some classical numerical characteristics of random variables and their distributions. The most popular ones are expected values and variances. More general characteristics are the *moments*. Among them, we emphasize moments about zero (about origin) and central moments.

Definition 1.7 For a discrete random variable X taking on values $x_1, x_2,$ ... with probabilities

$$p_k = P\{X = x_k\}, \qquad k = 1, 2, \ldots,$$

we define the nth *moment of X about zero* as

$$\alpha_n = EX^n = \sum_k x_k^n p_k. \tag{1.11}$$

We say that α_n exists if

$$\sum_k |x_k|^n p_k < \infty.$$

Note that the expected value EX is nothing but α_1. EX is also called the *mean of X* or the *mathematical expectation of X*.

Definition 1.8 The nth *central moment* of X is defined as

$$\beta_n = E(X - EX)^n = \sum_k (x_k - EX)^n p_k, \tag{1.12}$$

given that

$$\sum_k |x_k - EX|^n p_k < \infty.$$

If a random variable X has an absolutely continuous distribution with a pdf $p(x)$, then the moments about zero and the central moments have the following expressions:

$$\alpha_n = EX^n = \int_{-\infty}^{\infty} x^n p(x)\ dx \tag{1.13}$$

and

$$\beta_n = E(X - EX)^n = \int_{-\infty}^{\infty} (x - EX)^n p(x)\ dx. \tag{1.14}$$

We say that moments (1.13) exist if

$$\int_{-\infty}^{\infty} |x|^n p(x)\ dx < \infty. \tag{1.15}$$

The *variance of X* is simply the second central moment:

$$\text{Var } X = \beta_2 = E(X - EX)^2. \tag{1.16}$$

Central moments are easily expressed in terms of moments about zero as follows:

$$\begin{aligned}
\beta_n &= E(X - EX)^n = \sum_{k=0}^{n} (-1)^k \binom{n}{k}(EX)^k EX^{n-k} \\
&= \sum_{k=0}^{n} (-1)^k \binom{n}{k} \alpha_1^k \alpha_{n-k}.
\end{aligned} \tag{1.17}$$

In particular, we have

$$\text{Var } X = \beta_2 = \alpha_2 - \alpha_1^2 \tag{1.18}$$

and

$$\beta_3 = \alpha_3 - 3\alpha_1\alpha_2 + 2\alpha_1^3 \quad \text{and} \quad \beta_4 = \alpha_4 - 4\alpha_1\alpha_3 + 6\alpha_1^2\alpha_2 - 3\alpha_1^4. \tag{1.19}$$

Note that the first central moment $\beta_1 = 0$.

The inverse problem cannot be solved, however, because all central moments save no information about EX; hence, the expected value cannot be expressed in terms of β_n $(n = 1, 2, \ldots)$. Nevertheless, the relation

$$
\begin{aligned}
\alpha_n \quad = \quad & EX^n = E[(X - EX) + EX]^n \\
= \quad & \sum_{k=0}^{n} \binom{n}{k} (EX)^k E(X - EX)^{n-k} \\
= \quad & \sum_{k=0}^{n} \binom{n}{k} \alpha_1^k \beta_{n-k} \qquad\qquad (1.20)
\end{aligned}
$$

will enable us to express α_n $(n = 2, 3, \ldots)$ in terms of $\alpha_1 = EX$ and the central moments $\beta_2, \ldots, \beta_n$. In particular, we have

$$
\alpha_2 = \beta_2 + \alpha_1^2, \qquad\qquad (1.21)
$$

$$
\alpha_3 = \beta_3 + 3\beta_2\alpha_1 + \alpha_1^3 \quad \text{and} \quad \alpha_4 = \beta_4 + 4\beta_3\alpha_1 + 6\beta_2\alpha_1^2 + \alpha_1^4. \quad (1.22)
$$

Let X and Y belong to the same type of distribution [see (1.7)], meaning that

$$
Y \stackrel{d}{=} a + hX
$$

for some constants a and $h > 0$. Then, the following equalities allow us to express moments of Y in terms of the corresponding moments of X:

$$
EY^n = E(a + hX)^n = \sum_{k=0}^{n} \binom{n}{k} a^k h^{n-k} EX^{n-k} \qquad (1.23)
$$

and

$$
E(Y - EY)^n = E[h(X - EX)]^n = h^n E(X - EX)^n. \qquad (1.24)
$$

Note that the central moments of Y do not depend on the location parameter a. As particular cases of (1.23) and (1.24), we have

$$
\begin{aligned}
EY \quad &= \quad a + hEX, & (1.25) \\
EY^2 \quad &= \quad a^2 + 2ahEX + h^2 EX^2, \quad \text{Var } Y = h^2 \text{ Var } X, & (1.26) \\
EY^3 \quad &= \quad a^3 + 3a^2 hEX + 3ah^2 EX^2 + h^3 EX^3, & (1.27) \\
EY^4 \quad &= \quad a^4 + 4a^3 hEX + 6a^2 h^2 EX^2 + 4ah^3 EX^3 + h^4 EX^4. & (1.28)
\end{aligned}
$$

Definition 1.9 For random variables taking on values $0, 1, 2, \ldots$, the *factorial moments of positive order* are defined as

$$
\mu_r = EX(X - 1) \cdots (X - r + 1), \qquad r = 1, 2, \ldots, \qquad (1.29)
$$

while the *factorial moments of negative order* are defined as

$$
\mu_{-r} = E\left[\frac{1}{(X + 1)(X + 2) \cdots (X + r)} \right], \qquad r = 1, 2, \ldots. \qquad (1.30)
$$

While dealing with discrete distributions, it is quite often convenient to work with these factorial moments rather than regular moments. For this reason, it is useful to note the following relationships between the factorial moments and the moments:

$$\mu_1 = \alpha_1, \tag{1.31}$$

$$\mu_2 = \alpha_2 - \alpha_1, \tag{1.32}$$

$$\mu_3 = \alpha_3 - 3\alpha_2 + 2\alpha_1, \tag{1.33}$$

$$\mu_4 = \alpha_4 - 6\alpha_3 + 11\alpha_2 - 6\alpha_1, \tag{1.34}$$

$$\alpha_2 = \mu_2 + \mu_1, \tag{1.35}$$

$$\alpha_3 = \mu_3 + 3\mu_2 + \mu_1, \tag{1.36}$$

$$\alpha_4 = \mu_4 + 6\mu_3 + 7\mu_2 + \mu_1. \tag{1.37}$$

Exercise 1.4 Present two different random variables having the same expectations and the same variances.

Exercise 1.5 Let X be a random variable with expectation EX and variance Var X. What is the sign of $r(X) = E(X - |X|)(\text{Var } X - \text{Var } |X|)$? When does the quantity $r(X)$ equal 0?

Exercise 1.6 Suppose that X is a random variable such that $P\{X > 0\} = 1$ and that both EX and $E(1/X)$ exist. Then, show that $EX + E(1/X) \geq 2$.

Exercise 1.7 Suppose that $P\{0 \leq X \leq 1\} = 1$. Then, prove that $EX^2 \leq EX \leq EX^2 + \frac{1}{4}$. Also, find all distributions for which the left and right bounds are attained.

Exercise 1.8 Construct a variable X for which $EX^3 = -5$ and $EX^6 = 24$.

1.4   Shape Characteristics

For any distribution, we are often interested in some characteristics that are associated with the shape of the distribution. For example, we may be interested in finding out whether it is unimodal, or skewed, and so on. Two important measures in this respect are Pearson's measures of skewness and kurtosis.

Definition 1.10 Pearson's measures of skewness and kurtosis are given by

$$\gamma_1 = \frac{\beta_3}{\beta_2^{3/2}}$$

and

$$\gamma_2 = \frac{\beta_4}{\beta_2^2}.$$

Since these measures are functions of central moments, it is clear that they are free of the location. Similarly, due to the fractional form of the measures, it can readily be verified that they are free of scale as well. It can also be seen that the measure of skewness γ_1 may take on positive or negative values depending on whether β_3 is positive or negative, respectively. Obviously, when the distribution is symmetric about its mean, we may note that β_3 is 0, in which case the measure of skewness γ_1 is also 0. Hence, distributions with $\gamma_1 > 0$ are said to be *positively skewed distributions*, while those with $\gamma_1 < 0$ are said to be *negatively skewed distributions*.

Now, without loss of generality, let us consider an arbitrary distribution with mean 0 and variance 1. Then, by writing

$$\left[\int x^3 \, p(x) \, dx \right]^2 = \left[\int \left\{ x \sqrt{p(x)} \right\} \left\{ (x^2 - 1) \sqrt{p(x)} \right\} dx \right]^2$$

and applying the Cauchy–Schwarz inequality, we readily obtain the inequality

$$\gamma_2 \ge \gamma_1^2 + 1.$$

Later, we will observe the coefficient of kurtosis of a normal distribution to be 3. Based on this value, distributions with $\gamma_2 > 3$ are called *leptokurtic distributions*, while those with $\gamma_2 < 3$ are called *platykurtic distributions*. Incidentally, distributions for which $\gamma_2 = 3$ (which clearly includes the normal) are called *mesokurtic distributions*.

Remark 1.5 Karl Pearson (1895) designed a system of continuous distributions wherein the pdf of every member satisfies a differential equation. By studying their moment properties and, in particular, their coefficients of skewness and kurtosis, he proposed seven families of distributions which all occupied different regions of the (γ_1, γ_2)-plane. Several prominent distributions (such as beta, gamma, normal, and t that we will see in subsequent chapters) belong to these families. This development was the first and historic attempt to propose a unified mechanism for developing different families of statistical distributions.

1.5   Entropy

One more useful characteristic of distributions (called *entropy*) was introduced by Shannon.

Definition 1.11 For a discrete random variable X taking on values $x_1, x_2,$ $\ldots$ with probabilities $p_1, p_2, \ldots$, the *entropy* $H(X)$ is defined as

$$H(X) = -\sum_n p_n \log p_n. \tag{1.38}$$

If X has an absolutely continuous distribution with pdf $p(x)$, then the entropy is defined as

$$H(X) = -\int_D p(x) \log p(x) \, dx, \tag{1.39}$$

where

$$D = \{x : p(x) > 0\}.$$

In the case of discrete distributions, the transformation

$$Y = a + hX, \qquad -\infty < a < \infty, \; h > 0$$

does not change the probabilities p_n and, consequently, we have

$$H(Y) = H(X).$$

On the other hand, if X has a pdf $p(x)$, then $Y = a + hX$ has the pdf

$$g(x) = \frac{1}{h} p\left(\frac{x-a}{h}\right)$$

and

$$H(Y) = -\int_{D_1} g(x) \log g(x) \, dx,$$

where

$$D_1 = \{x : g(x) > 0\} = \left\{x : p\left(\frac{x-a}{h}\right) > 0\right\} = \left\{x : \frac{x-a}{h} \in D\right\}.$$

It is then easy to verify that

$$
\begin{aligned}
H(Y) &= -\int_{D_1} \frac{1}{h} p\left(\frac{x-a}{h}\right) \log\left\{\frac{1}{h} p\left(\frac{x-a}{h}\right)\right\} dx \\
&= -\int_D p(x) \log\left\{\frac{1}{h} p(x)\right\} dx \\
&= \log h \int_D p(x) \, dx - \int_D p(x) \log p(x) \, dx \\
&= \log h + H(X). \tag{1.40}
\end{aligned}
$$

1.6   Generating Function and Characteristic Function

In this section we present some functions that are useful in generating the probabilities or the moments of the distribution in a simple and unified manner. In addition, they may also help in identifying the distribution of an underlying random variable of interest.

Definition 1.12 Let X take on values $0, 1, 2, \ldots$ with probabilities $p_n = P\{X = n\}$, $n = 0, 1, \ldots$. All the information about this distribution is contained in the *generating function*, which is defined as

$$P(s) = Es^X = \sum_{n=0}^{\infty} p_n s^n, \tag{1.41}$$

with the right-hand side (RHS) of (1.41) converging at least for $|s| \leq 1$.

Some important properties of generating functions are as follows:

(a) $P(1) = 1$;

(b) for $|s| < 1$, there exist derivatives of $P(s)$ of any order;

(c) for $0 \leq s < 1$, $P(s)$ and all its derivatives $P^{(k)}(s), k = 1, 2, \ldots$, are nonnegative increasing convex functions;

(d) the generating function $P(s)$ uniquely determines probabilities p_n, $n = 1, 2, \ldots$, and the following relations are valid:

$$\begin{aligned} p_0 &= P(0), \\ p_n &= \frac{P^{(n)}(0)}{n!}, \qquad n = 1, 2, \ldots; \end{aligned}$$

(e) if random variables $X_1, \ldots, X_n$ are independent and have generating functions

$$P_k(s) = Es^{X_k}, \qquad k = 1, \ldots, n,$$

then the generating function of the sum $Y = X_1 + \cdots + X_n$ satisfies the relation

$$P_Y(s) = \prod_{k=1}^{n} P_k(s); \tag{1.42}$$

(f) the factorial moments can be determined from the generating function as

$$\mu_k = EX(X-1)\cdots(X-k+1) = P^{(k)}(1), \tag{1.43}$$

where

$$P^{(k)}(1) = \lim_{s \uparrow 1} P^{(k)}(s).$$

Definition 1.13 The *characteristic function* $f(t)$ of a random variable X is defined as

$$f(t) = E \exp\{itX\} = E \cos tX + i \, E \sin tX. \tag{1.44}$$

If X takes on values x_k ($k = 1, 2, \ldots$) with probabilities $p_k = P\{X = x_k\}$, then

$$
\begin{aligned}
f(t) &= \sum_k \exp(itx_k) \, p_k \\
&= \sum_k \cos(tx_k) \, p_k + i \sum_k \sin(tx_k) \, p_k.
\end{aligned} \tag{1.45}
$$

For a random variable having a pdf $p(x)$, the characteristic function takes on an analogous form:

$$
\begin{aligned}
f(t) &= \int_{-\infty}^{\infty} e^{itx} p(x) \, dx \\
&= \int_{-\infty}^{\infty} \cos(tx) \, p(x) \, dx + i \int_{-\infty}^{\infty} \sin(tx) \, p(x) \, dx.
\end{aligned} \tag{1.46}
$$

For random variables taking on values $0, 1, 2, \ldots$, there exists the following relationship between the characteristic function and the generating function:

$$f(t) = P(e^{it}). \tag{1.47}$$

Some of the useful properties of characteristic functions are as follows:

(a) $f(0) = 1$;

(b) $|f(t)| \leq 1$;

(c) $f(t)$ is uniformly continuous;

(d) $f(t)$ uniquely determines the distribution of the corresponding random variable X;

(e) if X has the characteristic function f, then $Y = a + hX$ has the characteristic function
$$g(t) = e^{iat} \, f(ht);$$

(f) if random variables $X_1, \ldots, X_n$ are independent and their characteristic functions are $f_1(t), \ldots, f_n(t)$, respectively, then the characteristic function of the sum $Y = X_1 + \cdots + X_n$ is given by

$$f_Y(t) = \prod_{k=1}^{n} f_k(t); \tag{1.48}$$

(g) if the nth moment EX^n of the random variable X exists, then the characteristic function $f(t)$ of X has the first n derivatives, and

$$\alpha_k = EX^k = \frac{f^{(k)}(0)}{i^k}\ , \qquad k = 1, 2, \ldots, n; \tag{1.49}$$

moreover, in this situation, the following expansion is valid for the characteristic function:

$$\begin{aligned} f(t) &= 1 + \sum_{k=1}^{n} f^{(k)}(0)t^k + r_n(t) \\ &= 1 + \sum_{k=1}^{n} \alpha_k (it)^k + r_n(t), \end{aligned} \tag{1.50}$$

where

$$r_n(t) = o(t^n)$$

as $t \to 0$;

(h) let random variables $X, X_1, X_2, \ldots$ have cdf's $F, F_1, F_2, \ldots$ and characteristic functions $f, f_1, f_2, \ldots$, respectively. If for any fixed t, as $n \to \infty$,

$$f_n(t) \to f(t), \tag{1.51}$$

then

$$F_n(x) \to F(x) \tag{1.52}$$

for any x, where the limiting cdf is continuous. Note that (1.52) also implies (1.51).

There exist inversion formulas for characteristic functions which will enable us to determine the distribution that corresponds to a certain characteristic function. For example, if

$$\int_{-\infty}^{\infty} |f(t)|\ dt < \infty,$$

where $f(t)$ is the characteristic function of a random variable X, then X has the pdf $p(x)$ given by

$$p(x) = \frac{1}{2\pi} \int_{-\infty}^{\infty} e^{-itx} f(t)\ dt. \tag{1.53}$$

Remark 1.6 Instead of working with characteristic functions, one could define the *moment generating function* of a random variable X as $E\exp\{tX\}$ (a real function this time) and work with it. However, there are instances where this moment generating function may not exist, while the characteristic function always exists. A classic example of this may be seen later when we discuss Cauchy distributions. Nonetheless, when the moment generating function does exist, it uniquely determines the distribution just as the characteristic function does.

Exercise 1.9 Consider a random variable X which takes on values $0, 1, 2, \ldots$ with probabilities $p_n = P\{X = n\}, n = 0, 1, 2, \ldots$. Let $P(s)$ be its generating function. If it is known that $P(0) = 0$ and $P(\frac{1}{3}) = \frac{1}{3}$, find the probabilities p_n.

Exercise 1.10 Let $P(s)$ and $Q(s)$ be the generating functions of the random variables X and Y. Suppose it is known that both EX and EY exist and that $P(s) \geq Q(s), 0 \leq s < 1$. What can be said about $E(X - Y)$? Can this expectation be positive, negative, or zero?

Exercise 1.11 If $f(t)$ is a characteristic function, then prove that the functions

$$f_1(t) = \frac{1}{2 - f(t)}, \quad f_2(t) = |f(t)|^2, \quad \text{and} \quad f_3(t) = \text{Re } f(t),$$

where Re $f(t)$ denotes the real part of $f(t)$, are also characteristic functions.

Exercise 1.12 If $f(t)$ is a characteristic function that is twice differentiable, prove that the function $g(t) = f''(t)/f''(0)$ is also a characteristic function.

Exercise 1.13 Consider the functions $f(t)$ and $g(t) = 2f(t) - 1$. Then, prove that if $g(t)$ is a characteristic function, $f(t)$ also ought to be a characteristic function. The reverse may not be true. To prove this, construct an example of a characteristic function $f(t)$ for which $g(t)$ is not a characteristic function.

Exercise 1.14 Find the only function among the following which is a characteristic function:

$$f(t), \quad f^2(2t), \quad f^3(3t), \quad \text{and} \quad f^6(6t).$$

Exercise 1.15 Find the only function among the following which is not a characteristic function:

$$f(t), \quad 2f(t) - 1, \quad 3f(t) - 2, \quad \text{and} \quad 4f(t) - 3.$$

Exercise 1.16 It is easy to verify that $f(t) = \cos t$ is a characteristic function of a random variable that takes on values 1 and -1 with equal probability of $\frac{1}{2}$. Consider now the following functions:

$$\cos^2 3t, \quad \cos^3 2t \cos^4 3t, \quad \cos t^2, \quad \cos(\cos t), \quad e^{\cos^3 t - 1} \quad \text{and} \quad \frac{1}{2 - \cos t}.$$

Which of these are characteristic functions?

Exercise 1.17 Prove that the functions $f_n(t) = \cos^n t - \sin^n t, n = 1, 2, \ldots$ are characteristic functions only if n is an even integer.

1.7   Decomposition of Distributions

Definition 1.14 We say that a random variable X is *decomposable* if there are two independent nondegenerate random variables Y and Z such that $X \overset{d}{=} Y + Z$.

Remark 1.7 An equivalent definition can be given in terms of characteristic functions as follows. A characteristic function f is *decomposable* if there are two nondegenerate characteristic functions f_1 anf f_2 such that

$$f(t) = f_1(t)f_2(t). \tag{1.54}$$

Note that degenerate characteristic functions have the form e^{ict}, which corresponds to the degenerate random variable taking on the value c with probability 1.

Definition 1.15 If for any $n = 1, 2, \ldots$, a characteristic function f satisfies the relation

$$f(t) = \{f_n(t)\}^n, \tag{1.55}$$

where $f_n(t)$ are characteristic functions for any $n = 1, 2, \ldots$, then we say that f is an *infinitely divisible characteristic function*. A random variable is said to have an *infinitely divisible distribution* if its characteristic function $f(t)$ is infinitely divisible.

Remark 1.8 Note that if a random variable has a bounded support, it cannot be infinitely divisible. It should also be noted that an infinitely divisible characteristic function cannot have zero values.

Exercise 1.18 Construct a decomposable random variable X for which X^2 is indecomposable.

Exercise 1.19 Construct two indecomposable random variables X and Y for which XY is decomposable.

1.8   Stable Distributions

Definition 1.16 We say that a charateristic function f is *stable* if, for any positive a_1 and a_2, there exist $a > 0$ and b such that

$$f(a_1 t)f(a_2 t) = e^{ibt} f(at). \tag{1.56}$$

A random variable is said to have a *stable distribution* if its characteristic function is stable.

Remark 1.9 It is of interest to note that any stable distribution is absolutely continuous, and is also infinitely divisible.

1.9   Random Vectors and Multivariate Distributions

Let $(\Omega, \mathcal{T}, P)$ be a probability space where $\Omega = \{\omega\}$ is a set of elementary events, $\mathcal{T}$ is a σ-algebra of events, and P is a probability measure defined on $(\Omega, \mathcal{T})$. Further, let B denote an element of the Borel σ-algebra of subsets of the n-dimensional Euclidean space $\mathcal{R}^n$.

Definition 1.17 An n-dimensional vector $\mathbf{X} = \mathbf{X}(\omega) = (\mathbf{X_1}(\omega), \ldots, \mathbf{X_n}(\omega))$ which maps Ω into $\mathcal{R}^n$ is called a *random vector* (or an n-dimensional random variable) if, for any Borel set $\mathbf{B}$ in $\mathcal{R}^n$, the inverse image of $\mathbf{B}$ given by

$$\mathbf{X}^{-1}(\mathbf{B}) = \{\omega : \mathbf{X}(\omega) \in \mathbf{B}\} = \{\omega : (\mathbf{X_1}(\omega), \ldots, \mathbf{X_n}(\omega)) \in \mathbf{B}\}$$

belongs to the σ-algebra $\mathcal{T}$.

This means that for any Borel set $\mathbf{B}$, we can define probability as

$$P\{\mathbf{X} \in \mathbf{B}\} = P\{\mathbf{X}^{-1}(\mathbf{B})\}.$$

In particular, for any $\mathbf{x} = (x_1, \ldots, x_n)$, the function

$$F(\mathbf{x}) = F(x_1, \ldots, x_n) \quad = \quad P\{X_1 \le x_1, \ldots, X_n \le x_n\},$$
$$-\infty < x_1, \ldots, x_n < \infty, \qquad (1.57)$$

is defined for the random vector $\mathbf{X} = (X_1, \ldots, X_n)$.

Definition 1.18 The function $F(\mathbf{x}) = F(x_1, \ldots, x_n)$ is called the *distribution function* of the random vector $\mathbf{X}$.

Remark 1.10 The elements $X_1, \ldots, X_n$ of the random vector $\mathbf{X}$ can be considered as n univariate random variables having distribution functions

$$F_1(x) \quad = \quad F(x, \infty, \ldots, \infty) = P\{X_1 \le x\},$$
$$F_2(x) \quad = \quad F(\infty, x, \infty, \ldots, \infty) = P\{X_2 \le x\},$$
$$\ldots$$
$$F_n(x) \quad = \quad F(\infty, \ldots, \infty, x) = P\{X_n \le x\},$$

respectively. Moreover, any set of n random variables $X_1, \ldots, X_n$ forms a random vector $\mathbf{X} = (X_1, \ldots, X_n)$. Hence,

$$F(\mathbf{x}) = F(x_1, \ldots, x_n) = P\{X_1 \le x_1, \ldots, X_n \le x_n\}$$

is often called the *joint distribution function* of the variables $X_1, \ldots, X_n$.

If $F(x_1, \ldots, x_n)$ is the joint distribution function of the random variables $X_1, \ldots, X_n$, we can obtain from it the joint distribution function of any subset $X_{\alpha(1)}, \ldots, X_{\alpha(m)}$ rather easily. For example, we have

$$P\{X_1 \le x_1, \ldots, X_m \le x_m\} = F(x_1, \ldots, x_m, \infty, \ldots, \infty) \qquad (1.58)$$

as the joint distribution function of $(X_1, \ldots, X_m)$.

Definition 1.19 The random variables $X_1, \ldots, X_n$ are said to be *independent random variables* if

$$P\{X_1 \le x_1, \ldots, X_n \le x_n\} = \prod_{k=1}^{n} P\{X_k \le x_k\} \qquad (1.59)$$

for any $-\infty < x_k < \infty$ $(k = 1, \ldots, n)$.

Definition 1.20 The vectors $\mathbf{X_1} = (X_1, \ldots, X_n)$ and $\mathbf{X_2} = (X_{n+1}, \ldots, X_{n+m})$ are said to be *independent* if

$$P\{X_1 \le x_1, \ldots, X_{n+m} \le x_{n+m}\}$$
$$= P\{X_1 \le x_1, \ldots, X_n \le x_n\} P\{X_{n+1} \le x_{n+1}, \ldots, X_{n+m} \le x_{n+m}\} \qquad (1.60)$$

for any $-\infty < x_k < \infty$ $(k = 1, \ldots, n + m)$.

In the following discussion we restrict ourselves to the two-dimensional case. Let (X, Y) be a two-dimensional random vector and let $F(x, y) = P(X \le x, Y \le y)$ be the joint distribution function of (X, Y). Then, $F_X(x) = F(x, \infty)$ and $F_Y(y) = F(\infty, y)$ are the marginal distribution functions. Now, as we did earlier in the univariate case, we shall discuss discrete and absolutely continuous cases separately.

Definition 1.21 A two-dimensional random vector (X, Y) is said to have a *discrete bivariate distribution* if there exists a countable set $B = \{(x_1, y_1), (x_2, y_2), \ldots\}$ such that $P\{(X, Y) \in B\} = 1$.

Remark 1.11 In order to determine a two-dimensional random vector (X, Y) having a bivariate discrete distribution, we need to fix two sequences: a sequence of two-dimensional points $(x_1, y_1), (x_2, y_2), \ldots$ and a sequence of probabilities $p_k = P\{X = x_k, Y = y_k\}, k = 1, 2, \ldots$, such that $\sum_k p_k = 1$. In this case, the joint distribution function $F(x, y)$ of (X, Y) is given by

$$F(x, y) = \sum_{k:\ x_k \le x, y_k \le y} p_k. \qquad (1.61)$$

Also, the components of the vector (X, Y) are independent if

$$P\{X = x_k, Y = y_k\} = P\{X = x_k\} P\{Y = y_k\} \quad \text{for any } k. \qquad (1.62)$$

Definition 1.22 A two-dimensional random vector (X, Y) with a joint distribution function $F(x, y)$ is said to have an *absolutely continuous bivariate distribution* if there exists a nonnegative function $p(u, v)$ such that

$$F(x, y) = \int_{-\infty}^{y} \int_{-\infty}^{x} p(u, v) \, du \, dv \qquad (1.63)$$

for any real x and y.

Remark 1.12 The function $p(u, v)$ satisfies the condition

$$\int_{-\infty}^{\infty} \int_{-\infty}^{\infty} p(u, v) \, du \, dv = 1, \tag{1.64}$$

and it is called the *probability density function* (pdf) of the bivariate random vector (X, Y) or the *joint probability density function* of the random variables X and Y. If $p(u, v)$ is the pdf of the bivariate vector (X, Y), then the components X and Y have one-dimensional (marginal) densities

$$p_X(u) = \int_{-\infty}^{\infty} p(u, v) \, dv \tag{1.65}$$

and

$$p_Y(v) = \int_{-\infty}^{\infty} p(u, v) \, du, \tag{1.66}$$

respectively.

Also, the components of the absolutely continuous bivariate vector (X, Y) are independent if

$$p(u, v) = p_X(u) p_Y(v), \tag{1.67}$$

where $p_X(u)$ and $p_Y(v)$ are the marginal densities as given in (1.65) and (1.66). Moreover, if the joint pdf $p(u, v)$ of (X, Y) admits a factorization of the form

$$p(u, v) = q_1(u) q_2(v), \tag{1.68}$$

then the components X and Y are independent, and there exists a nonzero constant c such that

$$p_X(u) = c \, q_1(u) \quad \text{and} \quad p_Y(v) = \frac{1}{c} \, q_2(v).$$

Exercise 1.20 Let $F(x, y)$ denote the distribution function of the random vector (X, Y). Then; express $P\{X \leq 0, Y > 1\}$ in terms of the function $F(x, y)$.

Exercise 1.21 Let $F(x)$ denote the distribution function of a random variable X. Consider the vector $\mathbf{X}_n = (X, \cdots, X)$ with all its components coinciding with X. Express the distribution function of $\mathbf{X}_n$ in terms of $F(x)$.

1.10   Conditional Distributions

Let (X, Y) be a random vector having a discrete bivariate distribution concentrated on some points (x_i, y_j), and let

$$p_{ij} = P\{X = x_i, Y = y_j\}, \; q_i = P\{X = x_i\} > 0, \; \text{and} \; r_j = P\{Y = y_j\} > 0,$$

for $i, j = 1, 2, \ldots$. Then, for any y_j $(j = 1, 2, \ldots)$, the *conditional distribution of X, given $Y = y_j$*, is defined as

$$P\{X = x_i | Y = y_j\} = \frac{P\{X = x_i, Y = y_j\}}{P\{Y = y_j\}} = \frac{p_{ij}}{r_j}. \qquad (1.69)$$

Similarly, for any x_i $(i = 1, 2, \ldots)$, the *conditional distribution of Y, given $X = x_i$*, is defined as

$$P\{Y = y_j | X = x_i\} = \frac{P\{X = x_i, Y = y_j\}}{P\{X = x_i\}} = \frac{p_{ij}}{q_i}. \qquad (1.70)$$

Next, let (X, Y) be an absolutely continuous bivariate random vector with a joint pdf $p(x, y)$ and marginal pdf's $p_X(x)$ and $p_Y(y)$. In this case, for any value y at which $p_Y(y)$ is continuous and positive, the *conditional pdf of X, given $Y = y$*, is defined as

$$p_{X|Y}(x|y) = \frac{p(x, y)}{p_Y(y)}. \qquad (1.71)$$

Similarly, for any value x at which $p_X(x)$ is continuous and positive, the *conditional pdf of Y, given $X = x$*, is defined as

$$p_{Y|X}(y|x) = \frac{p(x, y)}{p_X(x)}. \qquad (1.72)$$

Remark 1.13 Though the derivation of conditional distributions from a specified bivariate distribution is rather straightforward, the reverse is not so, however. The construction of a bivariate distribution with specified conditional distributions requires solutions of functional equations; for a detailed discussion, one may refer to the book by Arnold, Castillo, and Sarabia (1999).

In an analogous manner, we can also define conditional distributions for the general case of n-dimensional random vectors $(X_1, \ldots, X_n)$ with pdf $p_{\mathbf{X}}(x_1, \ldots, x_n)$. For example, let us consider the case when the n-dimensional random vector $\mathbf{X} = (X_1, \ldots, X_n)$ has an absolutely continuous distribution. Let $\mathbf{U} = (X_1, \ldots, X_m)$ and $\mathbf{V} = (X_{m+1}, \ldots, X_n)$ $(m < n)$ be the random vectors corresponding to the first m and the last $n - m$ components of the random vector $\mathbf{X}$. We can define the pdf's $p_{\mathbf{U}}(x_1, \ldots, x_m)$ and $p_{\mathbf{V}}(x_{m+1}, \ldots, x_n)$ in this case as in Eqs. (1.58) and (1.63). Then, the *conditional pdf of the random vector $\mathbf{V}$, given $\mathbf{U} = (x_1, \ldots, x_m)$*, is defined as

$$p_{\mathbf{V}|\mathbf{U}}(x_{m+1}, \ldots, x_n | x_1, \ldots, x_m) = \frac{p_{\mathbf{X}}(x_1, \ldots, x_n)}{p_{\mathbf{U}}(x_1, \ldots, x_m)} \qquad (1.73)$$

for any $(x_1, \ldots, x_m)$ at which $p_{\mathbf{U}}(x_1, \ldots, x_m) > 0$.

1.11   Moment Characteristics of Random Vectors

Let (X, Y) be a bivariate discrete random vector concentrating on points (x_i, y_j) with probabilities $p_{ij} = P\{X = x_i, Y = y_j\}$ for $i, j = 1, 2, \ldots$. For any measurable function $g(x, y)$, we can find the expected value of $Z = g(X, Y)$ as

$$EZ = Eg(X, Y) = \sum_i \sum_j g(x_i, y_j) p_{ij}. \tag{1.74}$$

Similarly, if (X, Y) has an absolutely continuous bivariate distribution with the density function $p(x, y)$, then we have the expected value of $Z = g(X, Y)$ as

$$EZ = Eg(X, Y) = \int_{-\infty}^{\infty} \int_{-\infty}^{\infty} g(x, y) p(x, y) \, dx \, dy. \tag{1.75}$$

Of course, as in the univariate case, we say that $EZ = Eg(X, Y)$ defined in Eqs. (1.74) and (1.75) exist if

$$\sum_i \sum_j |g(x_i, y_j)| \, p_{ij} < \infty \qquad \text{and} \qquad \int_{-\infty}^{\infty} \int_{-\infty}^{\infty} |g(x, y)| \, p(x, y) \, dx \, dy,$$

respectively. In particular, if $g(x, y) = x^k y^\ell$, we obtain $EZ = Eg(X, Y) = EX^k Y^\ell$, which is said to be the *product moment of order* (k, ℓ). Similarly, the moment $E(X - EX)^k (Y - EY)^\ell$ is said to be the *central product moment of order* (k, ℓ), and the special case of $E(X - EX)(Y - EY)$ is called the *covariance between X and Y* and is denoted by $\mathrm{Cov}(X, Y)$. Based on the covariance, we can define another measure of association which is invariant with respect to both location and scale of the variables X and Y (meaning that it is not affected if the means and the variances of the variables are changed). Such a measure is the *correlation coefficient between X and Y* and is defined as

$$\rho = \rho(X, Y) = \frac{\mathrm{Cov}(X, Y)}{\sqrt{\mathrm{Var}(X) \mathrm{Var}(Y)}}.$$

It can easily be shown that $|\rho| \leq 1$.

If we are dealing with a general n-dimensional random vector $\mathbf{X} = (X_1, \ldots, X_n)$, then the following moment characteristics of $\mathbf{X}$ will be of interest to us: the mean vector $\mathbf{m} = (m_1, \ldots, m_n)$, where $m_i = EX_i$ $(i = 1, \ldots, n)$, the covariance matrix $\Sigma = ((\sigma_{ij}))_{i,j=1}^n$, where $\sigma_{ij} = \sigma_{ji} = \mathrm{Cov}(X_i, X_j)$ (for $i \neq j$) and $\sigma_{ii} = \mathrm{Var}(X_i)$, and the correlation matrix $\rho = ((\rho_{ij}))_{i,j=1}^n$, where $\rho_{ij} = \sigma_{ij} / \sqrt{\sigma_{ii} \sigma_{jj}}$. Note that the diagonal elements of the correlation matrix are all 1.

Exercise 1.22 Find all distributions of the random variable X for which the correlation coefficient $\rho(X, X^2) = -1$.

Exercise 1.23 Suppose that the variances of the random variables X and Y are 1 and 4, respectively. Then, find the exact upper and lower bounds for Var $(X - Y)$.

1.12   Conditional Expectations

In Section 1.10 we introduced conditional distributions in the case of discrete as well as absolutely continuous multivariate distributions. Based on those conditional distributions, we describe in this section *conditional expectations*.

For this purpose, let us first consider the case when (X, Y) has a discrete bivariate distribution concentrating on points (x_i, y_j) (for $i, j = 1, 2, \ldots$), and as before, let $p_{ij} = P\{X = x_i, Y = y_j\}$ and $r_j = P\{Y = y_j\} > 0$. Suppose also that EX exists. Then, based on the definition of the conditional distribution of X, given $Y = y_j$, presented in Eq. (1.69), we readily have the *conditional mean of X, given $Y = y_j$*, as

$$E\left(X|Y = y_j\right) = \sum_i x_i P\{X = x_i|Y = y_j\} = \sum_i x_i \frac{p_{ij}}{r_j} . \qquad (1.76)$$

More generally, for any measurable function $h(\cdot)$ for which $Eh(X)$ exists, we have the *conditional expectation of $h(X)$, given $Y = y_j$*, as

$$E\left\{h(X)|Y = y_j\right\} = \sum_i h(x_i) P\{X = x_i|Y = y_j\} = \sum_i h(x_i) \frac{p_{ij}}{r_j}. \qquad (1.77)$$

Based on (1.76), we can introduce the *conditional expectation of X, given Y*, denoted by $E(X|Y)$, as a new random variable which takes on the value $E(X|Y = y_j)$ when Y takes on the value y_j (for $j = 1, 2, \ldots$). Hence, the conditional expectation of X, given Y, as a random variable takes on values

$$E\left(X|Y = y_j\right) = \sum_i x_i \frac{p_{ij}}{r_j}$$

with probabilities r_j (for $j = 1, 2, \ldots$). Consequently, we readily observe that

$$
\begin{aligned}
E\left\{E\left(X|Y\right)\right\} &= \sum_j E\left(X|Y = y_j\right) r_j \\
&= \sum_j r_j \sum_i x_i \frac{p_{ij}}{r_j} \\
&= \sum_i x_i \sum_j p_{ij} \\
&= \sum_i x_i P\{X = x_i\} = EX.
\end{aligned}
$$

Similarly, if the conditional expectation $E\{h(X)|Y\}$ is a random variable which takes on values $E\{h(X)|Y = y_j\}$ when Y takes on values y_j (for $j = 1, 2, \ldots$) with probabilities r_j, we can show that

$$E\left[E\{h(X)|Y\}\right] = E\{h(X)\}.$$

Next, let us consider the case when the random vector (X, Y) has an absolutely continuous bivariate distribution with pdf $p(x, y)$, and let $p_Y(y)$ be the marginal density function of Y. Then, from Eq. (1.71), we have the *conditional mean of X, given $Y = y$*, as

$$E(X|Y = y) = \int_{-\infty}^{\infty} x p_{X|Y}(x|y) \, dx = \int_{-\infty}^{\infty} x \, \frac{p(x, y)}{p_Y(y)} \, dx, \qquad (1.78)$$

provided that EX exists. Similarly, if $h(\cdot)$ is a measurable function for which $Eh(X)$ exists, we have the *conditional expectation of $h(X)$, given $Y = y$*, as

$$E\{h(X)|Y = y\} = \int_{-\infty}^{\infty} h(x) p_{X|Y}(x|y) \, dx = \int_{-\infty}^{\infty} h(x) \, \frac{p(x, y)}{p_Y(y)} \, dx. \quad (1.79)$$

As in the discrete case, we can regard $E(X|Y)$ and $E\{h(X)|Y\}$ as random variables which take on the values

$$E(X|Y = y) = \int_{-\infty}^{\infty} x \, \frac{p(x, y)}{p_Y(y)} \, dx, \ \ E\{h(X)|Y = y\} = \int_{-\infty}^{\infty} h(x) \, \frac{p(x, y)}{p_Y(y)} \, dx$$

when the random variable Y takes on the value y. In this case, too, it can be shown that

$$E\{E(X|Y)\} = EX \qquad \text{and} \qquad E\left[E\{h(X)|Y\}\right] = E\{h(X)\}.$$

1.13 Regressions

In Eqs. (1.76) and (1.78) we defined the conditional expectation $E(X|Y = y)$ provided that EX exists. From this conditional expectation, we may consider the function

$$a(y) = E(X|Y = y), \qquad (1.80)$$

which is called the *regression function of X on Y*. Similarly, when EY exists, the function

$$b(x) = E(Y|X = x) \qquad (1.81)$$

is called the *regression function of Y on X*.

Note that when the random variables X and Y are independent, then

$$a(y) = E(X|Y = y) = EX \qquad \text{and} \qquad b(x) = E(Y|X = x) = EY$$

are simply the unconditional means of X and Y, and do not depend on y and x, respectively.

1.14   Generating Function of Random Vectors

Let $\mathbf{X} = (X_1, \ldots, X_n)$ be a random vector, elements of which take on values $0, 1, 2, \ldots$. In this case, the *generating function* $P(s_1, \ldots, s_n)$ is defined as

$$
\begin{aligned}
P(s_1, \ldots, s_n) &= Es_1^{X_1} \cdots s_n^{X_n} \\
&= \sum_{j_1=0}^{\infty} \cdots \sum_{j_n=0}^{\infty} P\{X_1 = j_1, \ldots, X_n = j_n\}\, s_1^{j_1} \cdots s_n^{j_n}.
\end{aligned}
$$
(1.82)

Although the following properties can be presented for this general case, we present them for notational simplicity only for the bivariate case ($n = 2$).

Let $P_{X,Y}(s,t)$, $P_X(s)$, and $P_Y(t)$ be the generating function of the bivariate random vector (X, Y), the marginal generating function of X, and the marginal generating function of Y, defined by

$$
P_{X,Y}(s,t) = Es^X t^Y = \sum_{j=0}^{\infty} \sum_{k=0}^{\infty} P\{X = j, Y = k\} s^j t^k, \qquad (1.83)
$$

$$
P_X(s) = Es^X = \sum_{j=0}^{\infty} P\{X = j\} s^j, \qquad (1.84)
$$

$$
P_Y(t) = Et^Y = \sum_{k=0}^{\infty} P\{Y = k\} t^k, \qquad (1.85)
$$

respectively. Then, the following properties of $P_{X,Y}(s,t)$ can be established easily:

(a) $P_{X,Y}(1,1) = 1$;

(b) $P_{X,Y}(s,1) = P_X(s)$ and $P_{X,Y}(1,t) = P_Y(t)$;

(c) $P_{X,Y}(s,s) = P_{X+Y}(s)$, where $P_{X+Y}(s) = Es^{X+Y}$ is the generating function of the variable $X + Y$;

(d) $P_{X,Y}(s,t) = P_X(s)P_Y(t)$ if and only if X and Y are independent;

(e) $\left. \dfrac{\partial^{k+\ell} P_{X,Y}(s,t)}{\partial s^k \partial t^\ell} \right|_{s=t=1} = E\{X \cdots (X - k + 1)Y \cdots (Y - \ell + 1)\}$ and, in particular, we have

$$
\left. \frac{\partial^2 P_{X,Y}(s,t)}{\partial s\, \partial t} \right|_{s=t=1} = E\{XY\} .
$$

Next, for the random vector $\mathbf{X} = (X_1, \ldots, X_n)$, we define the *characteristic function* $f(t_1, \ldots, t_n)$ as

$$f(t_1,\ldots,t_n) = E\ e^{i(t_1 X_1 + \cdots + t_n X_n)}$$

$$= \sum_{j_1=0}^{\infty} \cdots \sum_{j_n=0}^{\infty} P\{X_1 = j_1, \ldots, X_n = j_n\} e^{i(t_1 j_1 + \cdots + t_n j_n)}.$$

$$(1.86)$$

Similarly, in the case when the random vector $\mathbf{X} = (X_1, \ldots, X_n)$ has an absolutely continuous distribution with density function $p(x_1, \ldots, x_n)$, then its *characteristic function* $f(t_1, \ldots, t_n)$ is defined as

$$f(t_1,\ldots,t_n) = E\ e^{i(t_1 X_1 + \cdots + t_n X_n)}$$

$$= \int_{-\infty}^{\infty} \cdots \int_{-\infty}^{\infty} e^{i(t_1 x_1 + \cdots + t_n x_n)} p(x_1, \ldots, x_n)\ dx_1 \cdots dx_n.$$

$$(1.87)$$

Once again, although the following properties can be presented for this general n-dimensional case, we present them for notational simplicity only for the bivariate case ($n = 2$).

Let $f_{X,Y}(s,t)$, $f_X(s)$, and $f_Y(t)$ be the characteristic function of the bivariate random vector (X,Y), the marginal characteristic function of X, and the marginal characteristic function of Y, defined by

$$f_{X,Y}(s,t) = \int_{-\infty}^{\infty}\int_{-\infty}^{\infty} e^{i(sx+ty)} p_{X,Y}(x,y)\ dx\ dy, \qquad (1.88)$$

$$f_X(s) = \int_{-\infty}^{\infty} e^{isx} p_X(x)\ dx, \qquad (1.89)$$

$$f_Y(t) = \int_{-\infty}^{\infty} e^{ity} p_Y(y)\ dy, \qquad (1.90)$$

respectively. Then, the following properties of $f_{X,Y}(s,t)$ can be established easily:

(a) $f_{X,Y}(0,0) = 1$;

(b) $f_{X,Y}(s,0) = f_X(s)$ and $f_{X,Y}(0,t) = f_Y(t)$;

(c) $f_{X,Y}(s,s) = f_{X+Y}(s)$, where $f_{X+Y}(s) = Ee^{is(X+Y)}$ is the characteristic function of the variable $X + Y$;

(d) $f_{X,Y}(s,t) = f_X(s)f_Y(t)$ if and only if X and Y are independent;

(e) $\left.\dfrac{\partial^{k+\ell} f_{X,Y}(s,t)}{\partial s^k \partial t^\ell}\right|_{s=t=0} = i^{k+\ell}\ E\left(X^k Y^\ell\right)$ and, in particular, we have

$$\left.\frac{\partial^2 f_{X,Y}(s,t)}{\partial s\ \partial t}\right|_{s=t=0} = -\ E(XY).$$

Exercise 1.24 Let $P(s,t)$ be the generating function of the random vector (X, Y). Then, find the generating function $Q(s, t, u)$ of the random vector $(2X + 1, X + Y, 3X + 2Y)$.

1.15 Transformations of Variables

Let $\mathbf{X} = (X_1, \ldots, X_n)$ be an n-dimensional random vector having a discrete multivariate distribution with a joint probability function $p(x_1, \ldots, x_n) = P\{X_1 = x_1, \ldots, X_n = x_n\}$ defined for all $(x_1, \ldots, x_n) \in B$, where $P\{\mathbf{X} \in B\} = 1$. Let $\mathbf{Y} = (Y_1, \ldots, Y_n) = (\phi_1(\mathbf{X}), \ldots, \phi_n(\mathbf{X}))$ define a one-to-one transformation that maps B onto C, where $P\{\mathbf{Y} \in C\} = 1$. Then, the joint probability function of the n-dimensional random vector $\mathbf{Y} = (Y_1, \ldots, Y_n) = (\phi_1(\mathbf{X}), \ldots, \phi_n(\mathbf{X}))$ is given by

$$
\begin{aligned}
q(\mathbf{y}) = q(y_1, \ldots, y_n) &= P\{Y_1 = y_1, \ldots, Y_n = y_n) \\
&= p(\psi_1(\mathbf{y}), \ldots, \psi_n(\mathbf{y})), \quad \mathbf{y} \in C, \quad (1.91)
\end{aligned}
$$

where $x_1 = \psi_1(y_1, \ldots, y_n), \ldots, x_n = \psi_n(y_1, \ldots, y_n)$ is the single-valued inverse of $y_1 = \phi_1(x_1, \ldots, x_n), \ldots, y_n = \phi_n(x_1, \ldots, x_n)$. Of course, the marginal probability function of any subset of the new variables may be obtained from (1.91) by summing over other variables.

Next, let us consider the case when $\mathbf{X} = (X_1, \ldots, X_n)$ is an n-dimensional random vector with an absolutely continuous distribution with density function $p(x_1, \ldots, x_n)$, and let B be a subset of $\mathcal{R}^n$. Now, consider the one-to-one transformation $y_1 = \phi_1(x_1, \ldots, x_n), \ldots, y_n = \phi_n(x_1, \ldots, x_n)$ which maps B onto C in the $(y_1, \ldots, y_n)$-space. Let its inverse transformation be

$$
x_1 = \psi_1(y_1, \ldots, y_n), \ldots, x_n = \psi_n(y_1, \ldots, y_n)
$$

with its first partial derivatives being continuous. Let the $n \times n$ determinant

$$
J = \begin{vmatrix}
\dfrac{\partial x_1}{\partial y_1} & \dfrac{\partial x_1}{\partial y_2} & \cdots & \dfrac{\partial x_1}{\partial y_n} \\
\dfrac{\partial x_2}{\partial y_1} & \dfrac{\partial x_2}{\partial y_2} & \cdots & \dfrac{\partial x_2}{\partial y_n} \\
\cdot & \cdot & \cdots & \cdot \\
\dfrac{\partial x_n}{\partial y_1} & \dfrac{\partial x_n}{\partial y_2} & \cdots & \dfrac{\partial x_n}{\partial y_n}
\end{vmatrix},
$$

called the *Jacobian of the transformation*, be not identically 0 in C. Then, the density function of the n-dimensional random vector $\mathbf{Y} = (Y_1, \ldots, Y_n) = (\phi_1(X_1, \ldots, X_n), \ldots, \phi_n(X_1, \ldots, X_n))$ is given by

$$
\begin{aligned}
q(\mathbf{y}) &= q_{Y_1, \ldots, Y_n}(y_1, \ldots, y_n) \\
&= p(\psi_1(y_1, \ldots, y_n), \ldots, \psi_n(y_1, \ldots, y_n)) |J|, \quad \text{for } \mathbf{y} \in C, \\
& \hspace{11cm} (1.92)
\end{aligned}
$$

where $|J|$ is the absolute value of the Jacobian of the transformation. Once again, the marginal pdf of any subset of the new variables may be obtained from (1.92) by integrating out the other variables.

Note that if the transformation is not one-to-one, but B is the union of a finite number of mutually disjoint spaces, say $B_1, \ldots, B_\ell$, then we can construct ℓ sets of one-to-one transformations (one for each B_i) and their respective Jacobians, and then finally express the density function of the vector $\mathbf{Y} = (Y_1, \ldots, Y_n)$ as the sum of ℓ terms of the form (1.92) corresponding to $B_1, \ldots, B_\ell$.

Part I

DISCRETE DISTRIBUTIONS

CHAPTER 2

DISCRETE UNIFORM DISTRIBUTION

2.1 Introduction

The general *discrete uniform distribution* takes on k distinct values $x_1, x_2,$ $\ldots, x_k$ with equal probabilities $1/k$, where k is a positive integer. We restrict our attention here to lattice distributions. In this case, $x_j = a + jh$, $j = 0, 1, \ldots, k-1$, where a is any real value and $h > 0$ is the step of the distribution. Sometimes, such a distribution is called a *discrete rectangular distribution*. The linear transformations of random variables enable us to consider, without loss of generality, just the standard discrete uniform distribution taking on values $0, 1, \ldots, k - 1$, which correspond to $a = 0$ and $h = 1$. Note that the case when $a = 0$ and $h = 1/k$ is also important, but it can be obtained from the standard discrete uniform distribution by means of a simple scale change.

2.2 Notations

We will use the notation

$$X \sim DU(k, a, h), \qquad -\infty < a < \infty, \ h > 0, \ k = 1, 2, \ldots ,$$

if

$$P\{X = a + jh\} = \frac{1}{k} \qquad \text{for } j = 0, 1, \ldots, k - 1,$$

and

$$X \sim DU(k)$$

for the corresponding standard discrete uniform distribution; i.e., $DU(k)$ is simply $DU(k, 0, 1)$.

Remark 2.1 Note that if

$$Y \sim DU(k, a, h) \quad \text{and} \quad X \sim DU(k),$$

then

$$X \stackrel{d}{=} \frac{Y - a}{h} \quad \text{and} \quad Y \stackrel{d}{=} a + hX,$$

where $\stackrel{d}{=}$ denotes "having the same distribution" (see Definition 1.5). More generally, if $Y_1 \sim DU(k, a_1, h_1)$ and $Y_2 \sim DU(k, a_2, h_2)$, then

$$Y_1 \stackrel{d}{=} cY_2 + d \quad \text{and} \quad Y_2 \stackrel{d}{=} \frac{Y_1 - d}{c},$$

where $c = h_1/h_2$ and $d = a_1 - a_2 h_1/h_2$. This means that the random variables Y_1 and Y_2 belong to the same type of distribution, depending only on the shape parameter k, and do not depend on location (a_1 and a_2) and scale (h_1 and h_2) parameters.

Discrete uniform distributions play a naturally important role in many classical problems of probability theory that deal with a random choice with equal probabilities from a finite set of k items. For example, a lottery machine contains k balls, numbered $1, 2, \ldots, k$. On selecting one of these balls, we get a random number Y which has the $DU(k, 1, 1)$ distribution. This principle, in fact, allows us to generate tables of random numbers used in different statistical simulations, by taking k sufficiently large (say, $k = 10^6$ or 2^{32}).

For the rest of this chapter, we deal only with the standard discrete uniform $DU(k)$ distribution.

2.3 Moments

We will now determine the moments of $X \sim DU(k)$, all of which exist since X has a finite support.

Moments about zero:

$$\alpha_n = EX^n = \frac{1}{k} \sum_{r=0}^{k-1} r^n, \qquad n = 1, 2, \ldots.$$

In particular,

$$\alpha_1 \;=\; EX = \frac{1}{k} \sum_{r=0}^{k-1} r = \frac{k-1}{2}, \tag{2.1}$$

$$\alpha_2 \;=\; EX^2 = \frac{1}{k} \sum_{r=0}^{k-1} r^2 = \frac{(k-1)(2k-1)}{6}, \tag{2.2}$$

$$\alpha_3 \;=\; EX^3 = \frac{1}{k} \sum_{r=0}^{k-1} r^3 = \frac{k(k-1)^2}{4}, \tag{2.3}$$

and

$$\alpha_4 = EX^4 = \frac{1}{k} \sum_{r=0}^{k-1} r^4 = \frac{(k-1)(2k-1)(3k^2 - 3k - 1)}{30}. \tag{2.4}$$

To obtain the expressions in (2.1)–(2.4), we have used the following well-known identities for sums:

$$\sum_{r=0}^{k-1} r = \frac{(k-1)k}{2}, \qquad \sum_{r=0}^{k-1} r^2 = \frac{(k-1)k(2k-1)}{6},$$

$$\sum_{r=0}^{k-1} r^3 = \frac{(k-1)^2 k^2}{4}, \quad \text{and} \quad \sum_{r=0}^{k-1} r^4 = \frac{(k-1)k(2k-1)(3k^2-3k-1)}{30};$$

see, for example, Gradshteyn and Ryzhik (1994, p. 2). Note that

$$\alpha_n \sim k^n/(n+1) \qquad \text{as } k \to \infty. \tag{2.5}$$

Central moments:

The variance or the second central moment is obtained from (2.1) and (2.2) as

$$\beta_2 = \text{Var } X = \alpha_2 - \alpha_1^2 = \frac{k^2 - 1}{12}. \tag{2.6}$$

The third central moment is obtained from (2.1)–(2.3) as

$$\begin{aligned}
\beta_3 = E(X - \alpha_1)^3 &= \alpha_3 - 3\alpha_2\alpha_1 + 2\alpha_1^3 \\
&= \frac{(k-1)^2}{4}\{k - (2k-1) + (k-1)\} = 0. \tag{2.7}
\end{aligned}$$

At first, (2.7) may seem surprising; but once we realize that X is symmetric about its mean value $\alpha_1 = (k-1)/2$, (2.7) makes perfect sense. In fact, it is easy to see that $(k - 1 - X)$ and X take on the same values $0, 1, \ldots, k-1$ with equal probabilities $1/k$. Therefore, we have

$$X - \alpha_1 \overset{d}{=} \alpha_1 - X,$$

and consequently,

$$\begin{aligned}
\beta_{2r+1} = E(X - \alpha_1)^{2r+1} &= E(\alpha_1 - X)^{2r+1} \\
&= (-1)^{2r+1} E(X - \alpha_1)^{2r+1} = -\beta_{2r+1},
\end{aligned}$$

which simply implies that

$$\beta_{2r+1} = 0, \qquad r = 1, 2, \ldots. \tag{2.8}$$

Factorial moments of positive order:

$$\begin{aligned}
\mu_r &= EX(X-1)\cdots(X-r+1) \\
&= \frac{1}{k}\sum_{m=0}^{k-1} m(m-1)\cdots(m-r+1), \qquad r = 1, 2, \ldots.
\end{aligned}$$

It is easily seen that $\mu_r = 0$ for $r \geq k$, and

$$
\begin{aligned}
\mu_r &= \frac{1}{k} \sum_{m=r}^{k-1} m(m-1)\cdots(m-r+1) \\
&= \frac{1}{k} \sum_{m=r}^{k-1} \frac{m!}{(m-r)!} \\
&= \frac{(k-1)(k-2)\cdots(k-r)}{r+1} \qquad \text{for } r = 1, 2, \ldots, k-1.
\end{aligned}
$$

In deriving the last expression, we have used the well-known combinatorial identity

$$
\sum_{m=r}^{k-1} \binom{m}{r} = \binom{k}{r+1}.
$$

In particular, we have

$$
\mu_1 = \alpha_1 = \frac{k-1}{2}, \tag{2.9}
$$

$$
\mu_2 = \alpha_2 - \alpha_1 = \frac{(k-1)(k-2)}{3}, \tag{2.10}
$$

$$
\mu_3 = \alpha_3 - 3\alpha_2 + 2\alpha_1 = \frac{(k-1)(k-2)(k-3)}{4}, \tag{2.11}
$$

and

$$
\mu_{k-2} = (k-2)!, \tag{2.12}
$$

$$
\mu_{k-1} = \frac{(k-1)!}{k}. \tag{2.13}
$$

Factorial moments of negative order:

$$
\begin{aligned}
\mu_{-r} &= E\left\{ \frac{1}{(X+1)\cdots(X+r)} \right\} \\
&= \frac{1}{k} \sum_{m=0}^{k-1} \frac{1}{(m+1)\cdots(m+r)}, \qquad r = 1, 2, \ldots.
\end{aligned}
$$

In particular,

$$
\mu_{-1} = \frac{1}{k} \sum_{m=1}^{k} \frac{1}{m} \tag{2.14}
$$

and

$$
\mu_{-1} \sim \frac{1}{k} \ln k \qquad \text{as } k \to \infty, \tag{2.15}
$$

$$
\mu_{-2} = \frac{1}{k} \sum_{m=0}^{k-1} \frac{1}{(m+1)(m+2)}
$$

$$= \frac{1}{k} \left(\sum_{m=0}^{k-1} \frac{1}{m+1} - \sum_{m=0}^{k-1} \frac{1}{m+2} \right)$$

$$= \frac{1}{k+1}, \tag{2.16}$$

$$\mu_{-3} = \frac{1}{k} \sum_{m=0}^{k-1} \frac{1}{(m+1)(m+2)(m+3)}$$

$$= \frac{1}{2k} \left(\sum_{m=0}^{k-1} \frac{1}{m+1} - 2 \sum_{m=0}^{k-1} \frac{1}{m+2} + \sum_{m=0}^{k-1} \frac{1}{m+3} \right)$$

$$= \frac{k+3}{4(k+1)(k+2)}. \tag{2.17}$$

2.4   Generating Function and Characteristic Function

The generating function of $DU(k)$ distribution exists for any s and is given by

$$P_X(s) = E s^X = \frac{1}{k} \sum_{r=0}^{k-1} s^r.$$

For $s \neq 1$, it can be rewritten in the form

$$P_X(s) = \frac{1 - s^k}{k(1 - s)}. \tag{2.18}$$

For any $k = 1, 2, \ldots$, $P_X(s)$ is a polynomial of $(k-1)$th degree, and it is not difficult to see that its roots coincide with

$$s_j = \exp(2\pi i j/k), \qquad j = 1, 2, \ldots, k-1 \text{ for } k > 1.$$

This readily gives us the following form of the generating function:

$$P_X(s) = \frac{1}{k} \prod_{j=1}^{k-1} (s - s_j) = \frac{1}{k} \prod_{j=1}^{k-1} \{ s - \exp(2\pi i j/k) \}. \tag{2.19}$$

Another form of the generating function exploits the *hypergeometric function*, defined by

$$_2F_1[a, b; c; x] = 1 + \frac{abx}{c1!} + \frac{a(a+1)b(b+1)x^2}{c(c+1)2!}$$

$$+ \frac{a(a+1)(a+2)b(b+1)(b+2)x^3}{c(c+1)(c+2)3!} + \cdots. \tag{2.20}$$

The generating function, in terms of the hypergeometric function, is given by

$$P_X(s) = {}_2F_1[-n+1, 1; -n+1; s]. \tag{2.21}$$

Since the characteristic function and the generating function for nonnegative integer-valued random variables satisfy the relation

$$f_X(t) = E \exp(itX) = P_X(e^{it}),$$

if we change s by e^{it} in Eqs. (2.18), (2.19), and (2.21), we obtain the corresponding expressions for the characteristic function $f_X(t)$. For example, from (2.18) we get

$$f_X(t) = \frac{1 - e^{itk}}{k(1 - e^{it})}. \tag{2.22}$$

2.5   Convolutions

Let us take two independent random variables, both having discrete uniform distributions, say, $X \sim DU(k)$ and $Y \sim DU(r)$, $k \geq r$ (without loss of any generality). Then, what can we say about the distribution of the sum $Z = X + Y$? The distribution of Z is called the *convolution* (or composition) of the two initial distributions.

Exercise 2.1 It is clear that $0 \leq Z \leq k + r - 2$. Consider the three different situations, and prove that $P\{Z \leq m\}$ is given by

(a) $\dfrac{(m+1)(m+2)}{2kr}$ if $0 \leq m \leq r - 1$,

(b) $\dfrac{2m - r + 3}{2k}$ if $r - 1 \leq m \leq k - 1$,

(c) $1 - \dfrac{(r + k - 2 - m)(r + k - 1 - m)}{2kr}$ if $k \leq m \leq k + r - 2$.

$$\tag{2.23}$$

From (2.23), we readily obtain

$$P\{Z = m\} = \begin{cases} \dfrac{m+1}{kr} & \text{if } 0 \leq m \leq r - 1, \\[2mm] \dfrac{1}{k} & \text{if } r \leq m \leq k - 1, \\[2mm] \dfrac{r + k - 1 - m}{kr} & \text{if } k \leq m \leq r + k - 2, \\[2mm] 0 & \text{otherwise.} \end{cases} \tag{2.24}$$

One can see now that $r = 1$ is the only case when the convolution of two discrete uniform $DU(k)$ and $DU(r)$ distributions leads to the same distribution. Note that in this situation $P\{Z = 0\} = 1$, which means that Z has a degenerate distribution. Nevertheless, it turns out that convolution of more general nondegenerate discrete uniform distributions may belong to the same set of distributions.

Exercise 2.2 Suppose that $Z \sim DU(r, 0, s)$ and $Y \sim DU(s)$, where $r = 2, 3, \ldots$, $s = 2, 3, \ldots$ and that Y and Z are independent random variables. Show then that $U = Y + Z \sim DU(rs)$.

Remark 2.2 It is easy to see that $Z \overset{d}{=} sX$, where $X \sim DU(r)$. Hence, we get another equivalent form of the statement given in Exercise 2.2 as follows. If $X \sim DU(r)$ and $Y \sim DU(s)$, then the sum $sX + Y$ has the discrete uniform $DU(sr)$ distribution. Moreover, due to the symmetry argument, we can immediately obtain that the sum $X + rY$ also has the same $DU(sr)$ distribution.

2.6 Decompositions

Decomposition is an operation which is inverse to convolution. We want to know if a certain random variable can be represented as a sum of at least two independent random variables (see Section 1.7). Of course, any random variable X can be rewritten as a trivial sum of two terms $a + (X - a)$, the first of which is a degenerate random variable, but we will solve a more interesting problem: Is it possible, for a certain random variable X, to find a pair of nondegenerate independent random variables Y and Z such that

$$X \overset{d}{=} Y + Z?$$

Consider $X \sim DU(k)$, where k is a compound number. Let $k = rs$, where $r \geq 2$ and $s \geq 2$ are integers. It follows from the statement of Exercise 2.2 that X is decomposable as a sum of two random variables, both having discrete uniform distributions. Moreover, we note from Remark 2.2 that we have at least two different options for decomposition of $DU(rs)$ if $r \neq s$.

Let k be a prime integer now. The simplest case is $k = 2$, when X takes on two values. Of course, in this situation X is indecomposable, because any nondegenerate random variable takes on at least two values and hence it is easy to see that any sum $Y + Z$ of independent nondegenerate random variables has at least three values. Now we can propose that $DU(3)$ distribution is decomposable. In fact, there are a lot of random variables, taking on values 0, 1, and 2 with probabilities p_0, p_1, and p_2, that can be presented as a sum $Y + Z$, where both Y and Z take on values 0 and 1, probably with different probabilities. However, it turns out that one can not decompose a random variable, taking three values, if the corresponding probabilities are equal ($p_0 = p_1 = p_2 = \frac{1}{3}$).

Exercise 2.3 Prove that $DU(3)$ distribution is indecomposable.

In the general case when k is any prime integer, by considering the corresponding generating function

$$P_X(s) = \frac{1}{k}\sum_{r=0}^{k-1}s^r,$$

we see that the problem of decomposition in this case is equivalent to the following problem: Is it possible to present $P_X(s)$ as a product of two polynomials with positive coefficients if k is a prime? The negative answer was given in Krasner and Ranulac (1937), and independently by Raikov (1937a). Summarizing all these, we have the following result.

Theorem 2.1 *The discrete uniform $DU(k)$ distribution (for $k > 1$) is indecomposable iff k is a prime integer.*

It is also evident that $X \sim DU(k)$ is not infinitely divisible when k is a prime number. Moreover, it is known that any distribution with a finite support cannot be infinitely divisible (see Remark 1.6). This means that any $DU(k)$ distribution is not infinitely divisible.

2.7   Entropy

From the definition of the entropy $H(X)$ in (1.38), it is clear that the entropy of any $DU(k, a, h)$ distribution depends only on k. If $X \sim DU(k)$, then

$$H(X) = \log k. \tag{2.25}$$

It is of interest to mention here that among all the random variables taking on at most k values, any random variable taking on distinct values $x_1, x_2, \ldots, x_k$ with probabilities $p_j = 1/k$ $(j = 1, 2, \ldots, k)$ has $\log k$ to be the maximum possible value for its entropy.

2.8   Relationships with Other Distributions

The discrete uniform distribution forms the basis for the derivation of many distributions. Here, we present some key connections of the discrete uniform distribution to some other distributions:

(a) We have already mentioned that $DU(1)$ distribution takes on the value 0 with probability 1 and, in fact, coincides with the degenerate distribution, which is discussed in Chapter 3.

(b) One more special case of discrete uniform distributions is $DU(2)$ distribution. If $X \sim DU(2)$, then it takes on values 0 and 1 with equal probability (of $\frac{1}{2}$). This belongs to the Bernoulli type of distribution discussed in Chapter 4.

(c) Let us consider a sequence of random variables $X_n \sim DU(n)$, $n = 1, 2, \dots$. Let $Y_n = X_n/n$. One can see that for any $n = 1, 2, \dots$,

$$Y_n \sim DU\left(n, 0, \frac{1}{n}\right)$$

and it takes on values $0, 1/n, 2/n, \dots, (n-1)/n$ with equal probabilities $1/n$. Let us try to find the limiting distribution for the sequence $Y_1, Y_2, \dots$. The simplest way is to consider the characteristic function. The characteristic function of Y_n is given by

$$g_{Y_n}(t) = E\exp(itY_n) = E\exp\{i(t/n)X_n\} = f_n(t/n),$$

where $f_n(t)$ is the characteristic function of $DU(n)$ distribution. Using (2.22), we readily find that

$$g_{Y_n}(t) = \frac{1 - e^{it}}{n(1 - e^{it/n})}. \tag{2.26}$$

It is not difficult to see now that for any fixed t,

$$g_{Y_n}(t) \to g(t) \qquad \text{as } n \to \infty,$$

where

$$g(t) = \frac{e^{it} - 1}{it} = \int_0^1 e^{itx}\, dx. \tag{2.27}$$

The RHS of (2.27) shows that $g(t)$ is the characteristic function of a continuous distribution with probability density function $p(x)$, which is equal to 1 if $0 < x < 1$ and equals 0 otherwise. The distribution with the given pdf is said to be uniform $U(0, 1)$, which is discussed in detail in Chapter 11. From the convergence of the corresponding characteristic functions, we can immediately conclude that the uniform $U(0, 1)$ distribution is the limit for the sequence of random variables $Y_n = X_n/n$, where $X_n \sim DU(n)$. Thus, we have constructed an important "bridge" between continuous uniform and discrete uniform distributions.

CHAPTER 3

DEGENERATE DISTRIBUTION

3.1 Introduction

Consider the $DU(k)$ distributed random variable X when $k = 1$. One can see that X takes on only one value $x_0 = 0$ with probability 1. It gives an example of the *degenerate distribution*. In the general case, the degenerate random variable takes on only one value, say c, with probability 1. In the sequel, $X \sim D(c)$ denotes a random variable having a degenerate distribution concentrated at the only point c, $-\infty < c < \infty$.

Degenerate distributions assume a special place in distribution theory. They can be included as a special case of many families of probability distributions, such as normal, geometric, Poisson, and binomial. For any sequence of random variables $X_1, X_2, \ldots$ and any arbitrary c, we can always choose sequences of normalizing constants α_n and β_n such that the limiting distribution of random variables

$$Z_n = (X_n - \alpha_n)/\beta_n \quad \text{as } n \to \infty$$

will become the degenerate $D(c)$ distribution.

A very important role in queueing theory is played by degenerate distributions. Kendall (1953), in his classification of queueing systems, has even reserved a special letter to denote systems with constant interarrival times or constant service times of customers. For example, $M/D/3$ means that a queueing system has three servers, all interarrival times are exponentially distributed, and the service of each customer requires a fixed nonrandom time. It should be mentioned that practically only degenerate (D), exponential (M), and Erlang (E) have their own letters in Kendall's classification of queueing systems.

3.2 Moments

Degenerate distributions have all its moments finite. Let $X \sim D(c)$ in the following discussion.

Moments about zero:

$$\alpha_n = EX^n = c^n, \qquad n = 1, 2, \dots. \tag{3.1}$$

In particular,

$$\alpha_1 = EX = c \tag{3.2}$$

and

$$\alpha_2 = EX^2 = c^2. \tag{3.3}$$

The variance is

$$\beta_2 = \text{Var } X = \alpha_2 - \alpha_1^2 = 0. \tag{3.4}$$

Note that (3.4) characterizes degenerate distributions, meaning that they are the only distributions having zero variance.

Characteristic function:

$$f_X(t) = Ee^{itX} = e^{itc}, \tag{3.5}$$

and, in particular,

$$f_X(t) = 1 \qquad \text{if } c = 0.$$

3.3   Independence

It turns out that any random variable X, having degenerate $D(c)$ distribution, is independent of any arbitrarily chosen random variable Y. For observing this, we must check that for any x and y,

$$P\{X \le x, \ Y \le y\} = P\{X \le x\}P\{Y \le y\}. \tag{3.6}$$

Equality (3.6) is evidently true if $x < c$, in which case

$$P\{X \le x, \ Y \le y\} = 0 \ \text{ and } \ P\{X \le x\} = 0.$$

If $x \ge c$, then

$$P\{X \le x\} = 1 \ \text{ and } \ P\{X \le x, \ Y \le y\} = P\{Y \le y\},$$

and we see that (3.6) is once again true.

Exercise 3.1 Let $Y = X$. Show that if X and Y are independent, then X has a degenerate distribution.

3.4   Convolution

It is clear that the convolution of two degenerate distributions is degenerate; that is, if $X \sim D(c_1)$ and $Y \sim D(c_2)$, then $X + Y \sim D(c_1 + c_2)$. Note also that if $X \sim D(c)$ and Y is an arbitrary random variable, then $X + Y$ belongs to the same type of distribution as Y.

Exercise 3.2 Let X_1 and X_2 be independent random variables having a common distribution. Prove that the equality

$$X_1 + X_2 \stackrel{d}{=} X_1 \tag{3.7}$$

implies that $X_1 \sim D(0)$.

Remark 3.1 We see that (3.7) characterizes the degenerate distribution concentrated at zero. If we take $X_1 + c$ instead of X_1 on the RHS of (3.7), we get a characterization of $D(c)$ distribution. Moreover, if $X_1, X_2, \ldots$ are independent and identically distributed random variables, then the equality

$$X_1 + \cdots + X_\ell \stackrel{d}{=} X_1 + \cdots + X_k + c, \qquad 1 \leq k < \ell,$$

gives a characterization of degenerate $D\left(\dfrac{c}{\ell - k}\right)$ distribution.

3.5   Decomposition

There is no doubt that any degenerate random variable can be presented only as a sum of degenerate random variables, but even this evident statement needs to be proved.

Exercise 3.3 Let X and Y be independent random variables, and let $X + Y$ have a degenerate distribution. Show then that both X and Y are degenerate.

It is interesting to observe that even such a simple distribution as a degenerate distribution possesses its own special properties and also assumes an important role among a very large family of distributions.

CHAPTER 4

BERNOULLI DISTRIBUTION

4.1 Introduction

The next simplest case after the degenerate random variable is one that takes on two values, say $x_1 < x_2$, with nonzero probabilities p_1 and p_2, respectively. The discrete uniform $DU(2)$ distribution is exactly double-valued. Let us recall that if $X \sim DU(2)$, then

$$P\{X = 0\} = P\{X = 1\} = \frac{1}{2}. \tag{4.1}$$

It is easy to give an example of such a random variable. Let X be the number of heads that have appeared after a single toss of a balanced coin. Of course, X can be 0 or 1 and it satisfies (4.1). Similarly, unbalanced coins result in distributions with

$$P\{X = 1\} = 1 - P\{X = 0\} = p, \tag{4.2}$$

where $0 < p < 1$. Indeed, a false coin with two tails or heads can serve as a model with $p = 0$ or $p = 1$ in (4.2), but in these situations X has degenerate $D(0)$ or $D(1)$ distributions.

The distribution defined in (4.2) is known as the *Bernoulli distribution*.

4.2 Notations

If X is defined by (4.2), we denote it by

$$X \sim Be(p).$$

Any random variable Y taking on two values $x_0 < x_1$ with probabilities

$$P\{Y = x_1\} = 1 - P\{Y = x_0\} = p \tag{4.3}$$

can clearly be represented as

$$Y = (x_1 - x_0)X + x_0,$$

43

where $X \sim Be(p)$. This means that random variables defined by (4.2) and (4.3) have the same type of distribution. Hence, X can be called a *random variable of the Bernoulli $Be(p)$ type*. Moreover, any random variable taking on two values with positive probabilities belongs to one of the Bernoulli types.

In what follows, we will deal with distributions satisfying (4.2).

4.3   Moments

Let $X \sim Be(p)$, $0 < p < 1$. Since X has a finite support, we can guarantee the existence of all its moments.

Moments about zero:

$$\alpha_n = EX^n = p, \qquad n = 1, 2, \dots. \tag{4.4}$$

Exercise 4.1 Let X have a nondegenerate distribution and

$$EX^2 = EX^3 = EX^4.$$

Show then that there exists a p, $0 < p < 1$, such that

$$X \sim Be(p).$$

Variance:

$$\beta_2 = \text{Var } X = \alpha_2 - \alpha_1^2 = p(1 - p). \tag{4.5}$$

It is clear that $0 < \beta_2 \le \frac{1}{4}$, and β_2 attains its maximum when $p = \frac{1}{2}$.

Central moments:
From (4.4), we readily find that

$$\beta_n = p(1 - p)\left\{(1 - p)^{n-1} + (-1)^n p^{n-1}\right\} \qquad \text{for } n \ge 2. \tag{4.6}$$

The expression of the variance in (4.5) follows from (4.6) if we set $n = 2$.

Measures of skewness and kurtosis:
From (4.6), we find Pearson's coefficient of skewness as

$$\gamma_1 = \frac{\beta_3}{\beta_2^{3/2}} = \frac{1 - 2p}{\{p(1 - p)\}^{1/2}}.$$

This expression for γ_1 readily reveals that the distribution is negatively skewed when $p > \frac{1}{2}$ and is positively skewed when $p < \frac{1}{2}$. Also, $\gamma_1 = 0$ only when $p = \frac{1}{2}$ (in this case, the distribution is also symmetric).

Similarly, we find Pearson's coefficient of kurtosis as

$$\gamma_2 = \frac{\beta_4}{\beta_2^2} = \frac{1}{p(1-p)} - 3.$$

This expression for γ_2 [due to the fact noted earlier that $\beta_2 = p(1-p) \leq \frac{1}{4}$] readily implies that $\gamma_2 \geq 1$ and that $\gamma_2 = 1$ only when $p = \frac{1}{2}$. Thus, in the case of $Be(\frac{1}{2})$ distribution, we do see that $\gamma_2 = \gamma_1^2 + 1$, which means that the inequality presented in Section 1.4 cannot be improved in general.

Entropy:
   It is easy to see that

$$H(X) = -p \log p - (1-p) \log(1-p). \tag{4.7}$$

Indeed, the maximal value of $H(X)$ is attained when $p = \frac{1}{2}$, in which case it equals 1.

Characteristic function:
   For $X \sim Be(p)$, $0 < p < 1$, the characteristic function is of the form

$$f_X(t) = (1-p) + pe^{it}. \tag{4.8}$$

As a special case, we can also find the characteristic function of a random variable Y, taking on values -1 and 1 with equal probabilities, since Y can be expressed as

$$Y = 2X - 1,$$

where $X \sim Be(\frac{1}{2})$. It is easy to see that

$$f_Y(t) = \cos t.$$

4.4   Convolutions

Let $X_1, X_2, \ldots$ be independent and identically distributed $Be(p)$ random variables, and let

$$Y_n = X_1 + \cdots + X_n.$$

Many methods are available to prove that

$$P\{Y_n = m\} = \binom{n}{m} p^m (1-p)^{n-m}, \qquad m = 0, 1, \ldots, n. \tag{4.9}$$

We will use here the characteristic function for this purpose. Let $g_n, f_1, \ldots, f_n$ be the characteristic functions of $Y_n, X_1, \ldots, X_n$, respectively. From (4.8), we have

$$f_k(t) = (1 - p + pe^{it}), \qquad k = 1, 2, \ldots, n.$$

Then, we get

$$
\begin{aligned}
g_n(t) &= \prod_{k=1}^{n} f_k(t) = (1 - p + pe^{it})^n \\
&= \sum_{m=0}^{n} \binom{n}{m} (pe^{it})^m (1-p)^{n-m} \\
&= \sum_{m=0}^{n} \binom{n}{m} p^m (1-p)^{n-m} e^{itm}.
\end{aligned}
\qquad (4.10)
$$

One can see that the sum on the RHS of (4.10) coincides with the characteristic function of a discrete random variable taking on values m $(m = 0, 1, \ldots, n)$ with probabilities

$$
p_m = \binom{n}{m} p^m (1-p)^{n-m}, \qquad m = 0, 1, \ldots, n.
$$

Thus, the probability distribution of Y_n is given by (4.9). This distribution, called the *binomial distribution*, is discussed in detail in Chapter 5.

Further, from Chapter 3, we already know that any $X \sim Be(p)$ is indecomposable.

4.5   Maximal Values

We have seen above that sums of Bernoulli random variables do not have the same type of distribution as their summands, but Bernoulli distributions are stable with respect to another operation.

Exercise 4.2 Let

$$
X_k \sim Be(p_k), \qquad k = 1, 2, \ldots, n,
$$

be independent random variables, and let

$$
M_n = \max\{X_1, \ldots, X_n\}.
$$

Show that
$$
M_n \sim Be(q_n),
$$
where $q_n = 1 - \prod_{k=1}^{n}(1 - p_k)$.

Quite often, Bernoulli random variables appear as random indicators of different events.

Example 4.1 Let us consider a random sequence $\alpha_1, \alpha_2, \ldots, \alpha_n$ of length n, which consists of zeros and ones. We suppose that $\alpha_1, \alpha_2, \ldots, \alpha_n$ are independent random variables taking on values 0 and 1 with probabilities r and $1 - r$, respectively. We say that a peak is present at point k ($k = 2, 3, \ldots, n - 1$) if $\alpha_{k-1} = \alpha_{k+1} = 0$ and $\alpha_k = 1$. Let N_k be the total number of peaks present in the sequence $\alpha_1, \alpha_2, \ldots, \alpha_n$. What is the expected value of N_n?

To find EN_n, let us introduce the events

$$A_k = \{\alpha_{k-1} = 0, \ \alpha_k = 1, \ \alpha_{k+1} = 0\}$$

and random indicators

$$X_k = 1\{A_k\}, \qquad k = 2, 3, \ldots, n - 1.$$

Note that $X_k = 1$ if A_k happens, and $X_k = 0$ otherwise.

In this case,

$$
\begin{aligned}
P\{X_k = 1\} &= 1 - P\{X_k = 0\} \\
&= P\{A_k\} \\
&= P\{\alpha_{k-1} = 0, \ \alpha_k = 1, \ \alpha_{k+1} = 0\} = (1 - r)r^2
\end{aligned}
$$

and

$$X_k \sim Be(p), \qquad k = 2, 3, \ldots, n - 1,$$

where $p = (1 - r)r^2$. Now, it is easy to see that

$$EX_k = (1 - r)r^2, \qquad k = 2, 3, \ldots, n - 1,$$

and

$$EN_k = E(X_2 + \cdots + X_{n-1}) = (n - 2)(1 - r)r^2.$$

In addition to the classical Bernoulli distributed random variables, there is one more class of Bernoulli distributions which is often encountered. These involve random variables $Y_1, Y_2, \ldots$, which take on values ± 1. Based on these random variables, the sum $S_k = Y_1 + \cdots + Y_n$ ($n = 1, 2, \ldots$) forms different discrete random walks on the integer-valued lattice and result in some interesting probability problems.

4.6 Relationships with Other Distributions

(a) We have shown that convolutions of Bernoulli distributions give rise to binomial distribution, which is discussed in detail in Chapter 5.

(b) Let $X_1, X_2, \ldots$ be independent $Be(p)$, $0 < p < 1$, random variables. Introduce a new random variable N as

$$N = \min\{j : X_{j+1} = 0\};$$

that is, N is simply the number of 1's in the sequence $X_1, X_2, \ldots$ that precede the first zero.

It is easy to see that N can take on values 0, 1, 2, ..., and its probability distribution is

$$
\begin{aligned}
P\{N = n\} &= P\{X_1 = 1,\ X_2 = 1, \ldots,\ X_n = 1,\ X_{n+1} = 0\} \\
&= P\{X_1 = 1\}P\{X_2 = 1\} \cdots P\{X_n = 1\}P\{X_{n+1} = 0\} \\
&= (1-p)p^n, \qquad n = 0, 1, \ldots.
\end{aligned}
$$

This distribution, called a *geometric distribution*, is discussed in detail in Chapter 6.

CHAPTER 5

BINOMIAL DISTRIBUTION

5.1 Introduction

As shown in Chapter 4, convolutions of Bernoulli distributions give rise to a new distribution that takes on a fixed set of integer values $0, 1, \ldots, n$ with probabilities

$$p_m = P\{X = m\} = \binom{n}{m} p^m (1 - p)^{n-m}, \quad m = 0, 1, \ldots, n, \tag{5.1}$$

where $0 \le p \le 1$ and $n = 1, 2, \ldots$. The parameter p in (5.1) can be equal to 0 or 1, but in these situations X has degenerate distributions, $D(0)$ and $D(n)$, respectively, which are naturally special cases of these distributions. The probability distribution in (5.1) is called a *binomial distribution*.

5.2 Notations

Let X have the distribution in (5.1). Then, we say that X is a binomially distributed random variable, and denote it by

$$X \sim B(n, p).$$

We know that linear transformations

$$Y = a + hX, \qquad -\infty < a < \infty, \ h > 0,$$

have the same type of distributions and, hence, we need to study only the standard distribution of the given type.

If X satisfies (5.1) and $Y = a + hX$, then Y takes on values $a, a+h, \ldots, a+ nh$ and

$$P\{Y = a + mh\} = \binom{n}{m} p^m (1 - p)^{n-m}, \quad m = 0, 1, \ldots, n. \tag{5.2}$$

We say that Y belongs to the binomial type of distribution, and denote it by

$$Y \sim B(n, p, a, h),$$

a and $h > 0$ being location and scale parameters, respectively.

5.3   Useful Representation

As we know from Chapter 4, binomial random variables can be expressed as sums of independent Bernoulli random variables. Let $Z_1, Z_2, \ldots$ be independent Bernoulli $Be(p)$ random variables and $X \sim B(n, p)$. Then the following equality holds for any $n = 1, 2, \ldots$:

$$X \stackrel{d}{=} Z_1 + Z_2 + \cdots + Z_n. \tag{5.3}$$

Due to (5.3), we can easily obtain generating function and characteristic function of binomial random variables from the corresponding expressions of Bernoulli random variables.

5.4   Generating Function and Characteristic Function

Let $X \sim B(n, p)$. It follows from (5.3) and the independence of Z_i's that the generating function of X is

$$
\begin{aligned}
P_X(s) &= Es^X = Es^{Z_1 + \cdots + Z_n} = Es^{Z_1} Es^{Z_2} \cdots Es^{Z_n} \\
&= (P_Z(s))^n = (1 - p + ps)^n,
\end{aligned} \tag{5.4}
$$

where $P_Z(s) = 1 - p + ps$ is the common generating function of the Bernoulli random variables $Z_1, Z_2, \ldots, Z_n$.

Let $f_X(t)$ be the characteristic function of X. From (5.4), we readily obtain

$$f_X(t) = P_X(e^{it}) = (1 - p + pe^{it})^n. \tag{5.5}$$

As a consequence, if $Y \sim B(n, p, a, h)$, then $Y \stackrel{d}{=} a + hX$ and

$$
\begin{aligned}
f_Y(t) &= Ee^{it(a+hX)} = e^{ita} Ee^{ithX} \\
&= e^{ita} f_X(th) = e^{iat}(1 - p + pe^{ith})^n.
\end{aligned} \tag{5.6}
$$

5.5   Moments

Equality (5.3) immediately yields

$$EX = E(Z_1 + \cdots + Z_n) = nEZ_1 = np \tag{5.7}$$

as well as

$$\text{Var } X = \text{Var } (Z_1 + \cdots + Z_n) = n \text{ Var } Z_1 = np(1 - p). \tag{5.8}$$

Other moments of X can be found by differentiating the generating function in (5.4).

Factorial moments of positive order:

$$\begin{aligned} \mu_k &= EX(X-1)\cdots(X-k+1) = P_X^{(k)}(1) \\ &= n(n-1)\cdots(n-k+1)p^k, \qquad k = 1,2,\ldots. \end{aligned} \tag{5.9}$$

In particular, we have

$$\begin{aligned} \mu_1 &= np, &(5.10) \\ \mu_2 &= n(n-1)p^2, &(5.11) \\ \mu_3 &= n(n-1)(n-2)p^3, &(5.12) \end{aligned}$$

and

$$\mu_n = n!\, p^n. \tag{5.13}$$

Note that

$$\mu_{n+1} = \mu_{n+2} = \cdots = 0. \tag{5.14}$$

Factorial moments of negative order:
For any $k = 1,2,\ldots,$

$$\mu_{-k} = E\left(\frac{1}{(X+1)\cdots(X+k)}\right).$$

Exercise 5.1 Show that we can use the equality

$$\mu_{-k} = P_X^{(-k)}(1), \qquad k = 1,2,\ldots,$$

under the assumption that we consider the derivative of a negative order $-k$ as the following integral:

$$P_X^{(-k)}(1) = \int_0^1 \int_0^{t_{k-1}} \cdots \int_0^{t_1} P_X(s)\, ds\, dt_1 \cdots dt_{k-1}. \tag{5.15}$$

Now we can use (5.15) to find μ_{-1} and μ_{-2} as follows:

$$\begin{aligned} \mu_{-1} &= E\left(\frac{1}{\xi+1}\right) = \int_0^1 P_X(s)\, ds \\ &= \int_0^1 (1-p+ps)^n\, ds \\ &= \frac{1-(1-p)^{n+1}}{(n+1)p}, \end{aligned} \tag{5.16}$$

$$\mu_{-2} = E\left(\frac{1}{(X+1)(X+2)}\right)$$

$$= \int_0^1 \int_0^t (1-p+ps)^n \, ds \, dt$$

$$= \frac{1-(1-p)^{n+1}\{1+(n+1)p\}}{(n+1)(n+2)p^2} . \tag{5.17}$$

Moments about zero:

Relations (5.10)–(5.13) readily imply that

$$\alpha_1 = EX = \mu_1 = np, \tag{5.18}$$

$$\alpha_2 = \mu_1 + \mu_2 = np + n(n-1)p^2 = np\{1+(n-1)p\}, \tag{5.19}$$

$$\alpha_3 = \mu_1 + 3\mu_2 + \mu_3$$
$$= np\{1+3(n-1)p+(n-1)(n-2)p^2\}, \tag{5.20}$$

and

$$\alpha_4 = \mu_1 + 7\mu_2 + 6\mu_3 + \mu_4$$
$$= np\{1+7(n-1)p+6(n-1)(n-2)p^2+(n-1)(n-2)(n-3)p^3\}. \tag{5.21}$$

Central moments:

From (5.18) and (5.19), we obtain [see also (5.8)] the variance as

$$\text{Var } X = \beta_2 = np(1-p). \tag{5.22}$$

Similarly, we find from (5.20) and (5.21) the third and fourth central moments as

$$\beta_3 = \alpha_3 - 3\alpha_2\alpha_1 + 2\alpha_1^3 = np(1-p)(1-2p) \tag{5.23}$$

and

$$\beta_4 = \alpha_4 - 4\alpha_3\alpha_1 + 6\alpha_2\alpha_1^2 - 3\alpha_1^4$$
$$= 3n^2p^2(1-p)^2 + np(1-p)(1-6p+6p^2). \tag{5.24}$$

Shape characteristics:

From (5.22)–(5.24), we find Pearson's coefficients of skewness and kurtosis as

$$\gamma_1 = \frac{\beta_3}{\beta_2^{3/2}} = \frac{1-2p}{\sqrt{np(1-p)}} \tag{5.25}$$

and

$$\gamma_2 = \frac{\beta_4}{\beta_2^2} = 3 + \frac{1-6p+6p^2}{np(1-p)}, \tag{5.26}$$

respectively. From (5.25), it is clear that the binomial distribution is negatively skewed when $p > \frac{1}{2}$ and is positively skewed when $p < \frac{1}{2}$. The coefficient of skewness is 0 when $p = \frac{1}{2}$ (in this case, the distribution is also symmetric). Equation (5.25) also reveals that the skewness decreases as n increases. Furthermore, we note from (5.26) that the binomial distribution may be leptokurtic, mesokurtic, or platykurtic, depending on the value of p. However, it is clear that γ_1 and γ_2 tend to 0 and 3, respectively, as n tends to ∞ (which are, as mentioned in Section 1.4, the coefficients of skewness and kurtosis of a normal distribution). Plots of the binomial mass function presented in Figures 5.1 and 5.2 reflect these properties.

Exercise 5.2 Show that the binomial $B(n, p)$ distribution is

leptokurtic for $p < \frac{1}{2}\left(1 - \frac{1}{\sqrt{3}}\right)$ or $p > \frac{1}{2}\left(1 + \frac{1}{\sqrt{3}}\right)$,

platykurtic for $\frac{1}{2}\left(1 - \frac{1}{\sqrt{3}}\right) < p < \frac{1}{2}\left(1 + \frac{1}{\sqrt{3}}\right)$,

and mesokurtic for $p = \frac{1}{2}\left(1 - \frac{1}{\sqrt{3}}\right)$ or $\frac{1}{2}\left(1 + \frac{1}{\sqrt{3}}\right)$.

5.6 Maximum Probabilities

Among all the binomial $B(n, p)$ probabilities,

$$p_m = \binom{n}{m} p^m (1 - p)^{n-m}, \qquad m = 0, 1, \ldots, n,$$

it will be of interest to find the maximum values. It appears that there are two different cases:

(a) If $m^* = (n + 1)p$ is an integer, then

$$p_{m^*-1} = p_{m^*} = \max_{0 \le m \le n} p_m.$$

(b) If m^* is not an integer, then $p_{[m^*]}$ is the only maximum in the sequence $p_0, \ldots, p_n$, where $[m^*]$ is the integer part of $m^* = (n + 1)p$.

Bin(10,0.25)

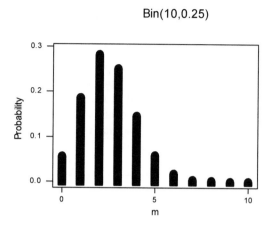

Bin(10,0.5)

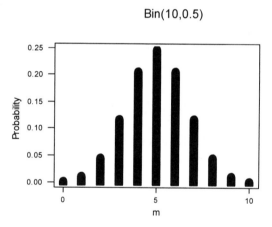

Bin(10,0.75)

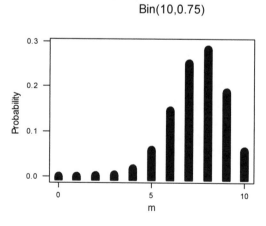

Figure 5.1. Plots of binomial mass function when $n = 10$

Bin(20,0.25)

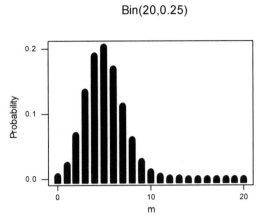

Bin(20,0.5)

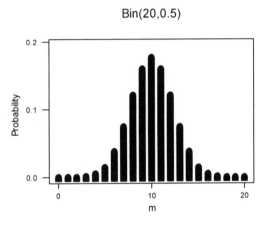

Bin(20,0.75)

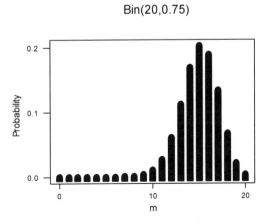

Figure 5.2. Plots of binomial mass function when $n = 20$

5.7   Convolutions

Let $X_1, X_2, \ldots$ be independent and identically distributed $B(n, p)$ random variables, and

$$S_N = X_1 + X_2 + \cdots + X_N, \qquad N = 1, 2, \ldots.$$

For finding the distribution of S_N, we may recall relation (5.3) to write

$$X_k \overset{d}{=} Z_{1k} + Z_{2k} + \cdots + Z_{nk}, \qquad k = 1, 2, \ldots, N, \qquad (5.27)$$

where all Z's in (5.27) are independent and have Bernoulli $Be(p)$ distribution. Then,

$$S_N \overset{d}{=} \sum_{k=1}^{N} \sum_{\ell=1}^{n} Z_{\ell k} \qquad (5.28)$$

is the sum of nN independent and identically distributed Bernoulli random variables, and therefore,

$$S_N \sim B(nN, p). \qquad (5.29)$$

Another way of establishing (5.29) is through characteristic functions.

Exercise 5.3 Using characteristic functions, show that the convolution of N binomial $B(n, p)$ distributions results in binomial $B(nN, p)$ distribution.

5.8   Decompositions

We know from (5.3) that any binomial $B(n, p)$ random variable can be expressed as a sum of n independent terms, each having a Bernoulli $Be(p)$ distribution. This simply means that any $B(n, p)$ distribution is decomposable for $0 < p < 1$ and $n \geq 2$. Moreover, $X \sim B(n, p)$ can be represented as a sum $U + V$ of two binomial random variables $U \sim B(m, p)$ and $V \sim B(n - m, p)$. Since $U + V = (U - a) + (V + a)$ for any $-\infty < a < \infty$, we can get for X a more general decomposition in terms of binomial distributions. In fact,

$$X \overset{d}{=} U + V$$

for any pair of independent random variables U and V, where

$$U \sim B(m, p, a, 1), \qquad V \sim B(n - m, p, -a, 1), \qquad (5.30)$$

describes a set of all possible decompositions of the binomial $B(n, p)$ distribution.

Exercise 5.4 Show that if $X \sim B(n, p)$, $n \geq 2$, $0 < p < 1$, then it can be decomposed into components of binomial type only as in (5.30).

Furthermore, since any binomial distribution has a finite support, it is not infinitely divisible.

5.9   Mixtures

Let us consider a sequence of binomial random variables $X_n \sim B(n, p)$, $n = 1, 2, \ldots$. Let one of these random variables be randomly selected. It means that we have a new random variable

$$Y = X_N,$$

where N is an integer-valued random variable which does not depend on X's. We are now interested in the distribution of Y. The distribution of Y is a mixture of binomial distributions since it can coincide with the distribution of the random variable X_n with probability $P\{N = n\}$. Of course, Y can take on values $0, 1, 2, \ldots$, and we want to find the probabilities $p_m = P\{Y = m\}$, $m = 0, 1, 2, \ldots$. These probabilities depend on the distribution of N. Suppose that $N \sim B(M, q)$ is also a binomial random variable. Since N can take on zero value, let us define $X_0 = 0$. Then, the theorem of total probability enables us to obtain the necessary probabilities as follows:

$$
\begin{aligned}
p_m &= P\{Y = m\} = P\{X_N = m\} \\
&= \sum_{n=m}^{M} P\{X_N = m | N = n\} P\{N = n\} \\
&= \sum_{n=m}^{M} P\{X_n = m | N = n\} P\{N = n\} \\
&= \sum_{n=m}^{M} P\{X_n = m\} P\{N = n\}.
\end{aligned}
\tag{5.31}
$$

In (5.31), the independence of X's and N has been used. As the distributions of X's and N are known to us, we find that

$$
\begin{aligned}
p_m &= \sum_{n=m}^{M} \frac{n!}{m!\,(n-m)!} p^m (1-p)^{n-m} \, \frac{M!}{n!\,(M-n)!} q^n (1-q)^{M-n} \\
&= \frac{M!}{m!} (pq)^m \sum_{n=m}^{M} \frac{\{(1-p)q\}^{n-m}(1-q)^{M-n}}{(n-m)!\,(M-n)!}
\end{aligned}
$$

$$
= \binom{M}{m}(pq)^m \sum_{n=0}^{M-m} \binom{M-m}{n}\{(1-p)q\}^n(1-q)^{M-m-n}
$$

$$
= \binom{M}{m}(pq)^m \{1-q+(1-p)q\}^{M-m} = \binom{M}{m}(pq)^m(1-pq)^{M-m}.
$$

$$(5.32)$$

It follows from (5.32) that $Y \sim B(M,pq)$.

5.10   Conditional Probabilities

Consider the following situation. Let $X_1 \sim B(n,p)$ and $X_2 \sim B(m,p)$ be independent random variables. Then, we have already seen that $Y = X_1 + X_2$ has a binomial $B(m+n,p)$ distribution. Suppose it is known that Y has some fixed value r, $0 \le r \le m+n$. Then, we are interested in finding the conditional distribution of X_1, given that $Y = r$. For this, we need to find the conditional probabilities $P\{X_1 = t | Y = r\}$ for all integers t in the interval

$$
\max(0, r-m) \le t \le \min(n, r). \tag{5.33}
$$

For example, if $r = n + m$, we must consider only the value $t = n$ and get that

$$
P\{X_1 = n | X_1 + X_2 = n + m\} = 1.
$$

Now, for any t satisfying (5.33), we find that

$$
\begin{aligned}
P\{X_1 = t | Y = r\} &= \frac{P\{X_1 = t, Y = r\}}{P\{Y = r\}} \\
&= \frac{P\{X_1 = t\}P\{X_2 = r - t\}}{P\{Y = r\}} \\
&= \frac{\binom{n}{t}p^t(1-p)^{n-t}\binom{m}{r-t}p^{r-t}(1-p)^{m-r+t}}{\binom{n+m}{r}p^r(1-p)^{n+m-r}} \\
&= \frac{\binom{n}{t}\binom{m}{r-t}}{\binom{n+m}{r}}. \tag{5.34}
\end{aligned}
$$

It is curious that the RHS of (5.34) does not depend on p. Let us now consider the following classical problem from combinatorics: A box contains n red and m black balls. A random selection of r balls is made from this box. What is the probability of getting exactly t red balls in our sample? It is easy to see that this probability coincides with the RHS of (5.34). To continue the problem, introduce a random variable N for the number of red balls taken from the box. Indeed, N takes on the values in (5.33) with probabilities as in (5.34). Thus, the conditional distribution of X_1, given that $X_1 + X_2$ is fixed, coincides with the distribution of N, called a *hypergeometric distribution*, which is discussed in detail in Chapter 8.

5.11   Tail Probabilities

Let $X \sim B(n, p)$. Quite often, one has to calculate probabilities $P\{X < m\}$ or $P\{X \geq m\}$. Of course, we can write that

$$P\{X \geq m\} = 1 - P\{X < m\} = \sum_{k=m}^{n} \binom{n}{m} p^m (1 - p)^{n-m}, \qquad (5.35)$$

but the computation of the RHS of (5.35) may become difficult for large values of m and n. For this purpose, a simpler integration formula can be given.

Exercise 5.5 Prove that

$$\sum_{k=m}^{n} \binom{n}{k} p^k (1 - p)^{n-k} = \frac{n!}{(m - 1)! \, (n - m)!} \int_0^p x^{m-1} (1 - x)^{n-m} \, dx.$$

$$(5.36)$$

We will recall equality (5.36) in Chapter 16 when we discuss the *beta distribution*.

5.12   Limiting Distributions

There are two important distributions which appear as the limit for some sequences of binomial distributions.

(a) For a fixed positive λ and $n > \lambda$, consider a sequence of random variables $X_n \sim B(n, \lambda/n)$. In this case

$$P\{X_n = m\} = \binom{n}{m} \left(\frac{\lambda}{n}\right)^m \left(1 - \frac{\lambda}{n}\right)^{n-m}, \qquad m = 0, 1, \ldots, n.$$

Exercise 5.6 Show that for any fixed $m = 0, 1, \ldots,$

$$\lim_{n \to \infty} P\{X_n = m\} = p_m(\lambda), \tag{5.37}$$

where

$$p_m(\lambda) = e^{-\lambda} \frac{\lambda^m}{m!}, \qquad m = 0, 1, \ldots. \tag{5.38}$$

Note that a random variable taking on values $0, 1, 2, \ldots$ with probabilities $p_m(\lambda)$ is said to have a *Poisson distribution*, and it is discussed in detail in Chapter 9.

(b) Let $X_n \sim B(n, p)$, $n = 1, 2, \ldots$. We know that

$$EX_n = np \qquad \text{and} \qquad \text{Var } X_n = np(1 - p).$$

Consider a new sequence of normalized random variables

$$W_n = \frac{X_n - EX_n}{\sqrt{\text{Var } X_n}} = \frac{X_n - np}{\sqrt{np(1 - p)}}, \qquad n = 1, 2, \ldots. \tag{5.39}$$

Using notations introduced in Section 5.2, we can write

$$W_n \sim B(n, p, a_n, h_n),$$

where

$$a_n = -\sqrt{\frac{np}{1 - p}}$$

and

$$h_n = \frac{1}{\sqrt{np(1 - p)}}.$$

For the new random variables, we readily find that

$$EW_n = 0 \qquad \text{and} \qquad \text{Var } W_n = 1, \qquad n = 1, 2, \ldots.$$

Let $f_n(t)$ and $g_n(t)$ be the characteristic functions of X_n and W_n, respectively. From (5.5) and (5.6), it follows that

$$f_n(t) = (1 - p + pe^{it})^n$$

and

$$\begin{aligned}
g_n(t) &= \exp\{ia_n t\} f_n(h_n t) \\
&= \exp\left\{ -it\sqrt{\frac{np}{1 - p}} \right\} \left(1 - p + p \exp\left\{ \frac{it}{\sqrt{np(1 - p)}} \right\} \right)^n.
\end{aligned} \tag{5.40}$$

From (5.40), we obtain that for any fixed t,

$$g_n(t) \to e^{-t^2/2} \text{ as } n \to \infty. \tag{5.41}$$

Since

$$e^{-t^2/2} = \int_{-\infty}^{\infty} e^{itx} \frac{1}{\sqrt{2\pi}} e^{-x^2/2} dx \tag{5.42}$$

is the characteristic function of the density function

$$p(x) = \frac{1}{\sqrt{2\pi}} e^{-x^2/2}, \tag{5.43}$$

we conclude that the sequence of distributions of W_n converges as $n \to \infty$ to a limiting distribution with density function $p(x)$ as in (5.43). This continuous distribution, called the *standard normal distribution*, is discussed in detail in Chapter 23.

Thus, we have investigated the asymptotic behavior of two sequences of binomial random variables and obtained two different limiting distributions, Poisson and normal.

CHAPTER 6

GEOMETRIC DISTRIBUTION

6.1 Introduction

Consider the following scenario. There is a sequence of independent trials in each of which a certain event A can occur with a constant probability p, and its complement $\bar{A}$ can occur with probability $1 - p$. Let us call A and $\bar{A}$ "success" and "failure", respectively. Now, let X_k $(k = 1, 2, \ldots)$ be the number of "successes" in the kth trial. Indeed, for any k, $X_k = \mathbf{1}_{\{A\}}$ and X_k takes on two values, 0 and 1, with probabilities $1 - p$ and p, respectively. This means that $X_k \sim Be(p)$, $k = 1, 2, \ldots$. Let Y_n be the total number of "successes" in the first n trials; i.e., $Y_n = X_1 + X_2 + \cdots + X_n$. We know that the convolution of Bernoulli $Be(p)$ random variables generates the binomial $B(n, p)$ distribution. Hence, $Y_n \sim B(n, p)$ for any $n = 1, 2, \ldots$. Now, let N be the number of "successes" until the appearance of the first "failure". Then, this example with a sequence of ones and zeros considered in Chapter 4 revealed that

$$P\{N = m\} = (1 - p)p^m, \qquad m = 0, 1, \ldots.$$

This probability distribution is the subject matter of this chapter.

6.2 Notations

Let a random variable X take on values $0, 1, 2, \ldots$ with probabilities

$$p_n = P\{X = n\} = (1 - p)p^n, \qquad n = 0, 1, \ldots, \tag{6.1}$$

where $0 < p < 1$. We then say that X has a *geometric distribution* with parameter p, and denote it by

$$X \sim G(p).$$

The geometric type of distributions include all linear transformations $Y = a + hX$, $-\infty < a < \infty$ and $h > 0$, of the geometric random variable. In this case we denote it by $Y \sim G(p, a, h)$, which is a random variable concentrated at points $a, a + h, a + 2h, \ldots$ with probabilities

$$P\{Y = a + nh\} = (1 - p)p^n, \qquad n = 0, 1, \ldots. \tag{6.2}$$

6.3   Tail Probabilities

If $X \sim G(p)$, then the tail probabilities of X are very simple. In fact,

$$P\{X \geq n\} = \sum_{k=n}^{\infty}(1-p)p^k = p^n, \qquad n = 0, 1, 2, \ldots. \qquad (6.3)$$

Formula (6.3) can be used to solve the following problem.

Exercise 6.1 Let $X_k \sim G(p_k)$, $k = 1, 2, \ldots, n$, be a sequence of independent random variables, and

$$m_n = \min_{1 \leq k \leq n} \xi_k.$$

Prove that $m_n \sim G(p)$, where $p = \prod_{k=1}^{n} p_k$.

6.4   Generating Function and Characteristic Function

If $X \sim G(p)$, then

$$P_X(s) = Es^X = \sum_{k=0}^{\infty}(1-p)p^k s^k = \frac{1-p}{1-ps} \qquad \text{if } |s| < \frac{1}{p} \qquad (6.4)$$

and

$$f_X(t) = Ee^{itX} = P_X(e^{it}) = \frac{1-p}{1-pe^{it}}. \qquad (6.5)$$

6.5   Moments

The generating function in (6.4) has a rather simple form which we will use first to derive expressions for factorial moments.

Factorial moments of positive order:

$$\mu_k = EX(X-1)\cdots(X-k+1) = P_X^{(k)}(1)$$
$$= \frac{k!\, p^k}{(1-p)^k}, \qquad k = 1, 2, \ldots. \ (6.6)$$

In particular, we have

$$\mu_1 = \frac{p}{1-p}, \tag{6.7}$$

$$\mu_2 = \frac{2p^2}{(1-p)^2}, \tag{6.8}$$

$$\mu_3 = \frac{6p^3}{(1-p)^3}, \tag{6.9}$$

and

$$\mu_4 = \frac{24p^4}{(1-p)^4}. \tag{6.10}$$

Factorial moments of negative order:
From formula (5.15) we can give the general form of moments:

$$\mu_{-k} = E\left\{\frac{1}{(X+1)\cdots(X+k)}\right\}, \qquad k = 1, 2, \ldots,$$

which is as follows:

$$
\begin{aligned}
\mu_{-k} &= \int_0^1 \int_0^{t_{k-1}} \cdots \int_0^{t_1} P_X(s)\, ds\, dt_1 \cdots dt_{k-1} \\
&= (1-p) \int_0^1 \int_0^{t_{k-1}} \cdots \int_0^{t_1} \frac{ds}{1-ps}\, dt_1 \cdots dt_{k-1}. \tag{6.11}
\end{aligned}
$$

Now,

$$\mu_{-1} = (1-p)\int_0^1 \frac{ds}{1-ps} = -\frac{(1-p)\ln(1-p)}{p} \tag{6.12}$$

and

$$
\begin{aligned}
\mu_{-2} &= (1-p)\int_0^1 \int_0^t \frac{ds}{1-ps}\, dt \\
&= \frac{1-p}{p} + \frac{(1-p)^2}{p^2}\ln(1-p). \tag{6.13}
\end{aligned}
$$

Moments about zero:
It follows from (6.7)–(6.10) that

$$\alpha_1 = EX = \mu_1 = \frac{p}{1-p}, \tag{6.14}$$

$$\alpha_2 = EX^2 = \mu_1 + \mu_2 = \frac{p(1+p)}{(1-p)^2}, \tag{6.15}$$

$$\alpha_3 = EX^3 = \mu_1 + 3\mu_2 + \mu_3 = \frac{p(1+4p+p^2)}{(1-p)^3}, \tag{6.16}$$

and

$$\alpha_4 = EX^4 = \mu_1 + 7\mu_2 + 6\mu_3 + \mu_4 = \frac{p(1+p)(1+10p+p^2)}{(1-p)^4} . \qquad (6.17)$$

Central moments:
   From (6.14) and (6.15), we readily obtain the variance to be

$$\beta_2 = \text{Var } X = \alpha_2 - \alpha_1^2 = \frac{p}{(1-p)^2} . \qquad (6.18)$$

   Similarly, from (6.14)–(6.17), we obtain the third and fourth central moments to be

$$\beta_3 = \frac{p(1+p)}{(1-p)^3} \qquad (6.19)$$

and

$$\beta_4 = \frac{p(1+7p+p^2)}{(1-p)^4} , \qquad (6.20)$$

respectively.

Shape characteristics:
   From (6.18)–(6.20), we find Pearson's coefficients of skewness and kurtosis as

$$\gamma_1 = \frac{\beta_3}{\beta_2^{3/2}} = \frac{1+p}{\sqrt{p}} \qquad (6.21)$$

and

$$\gamma_2 = \frac{\beta_4}{\beta_2^2} = \frac{1+7p+p^2}{p} , \qquad (6.22)$$

respectively. It is quite clear from (6.21) that the geometric distribution is positively skewed. Furthermore, we observe from (6.22) that the distribution is leptokurtic for all values of the parameter p. Figure 6.1 presents plots of the geometric mass function for some choices of p.

Exercise 6.2 Prove that the geometric $G(p)$ distribution is leptokurtic for all values of the parameter p.

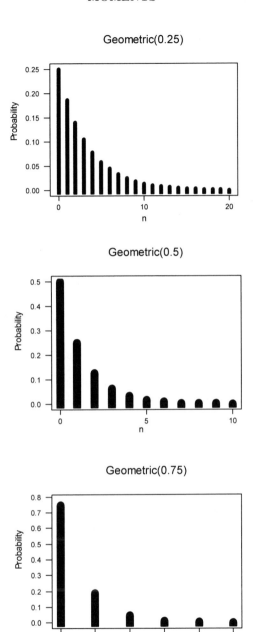

Figure 6.1. Plots of geometric mass function

6.6 Convolutions

Let $X_1 \sim G(p_1)$ and $X_2 \sim G(p_2)$ be independent random variables. Without loss of generality, let us assume that $p_1 > p_2$. We are then interested in the distribution of $Y = X_1 + X_2$; see Sen and Balakrishnan (1999) for a discussion on a more general problem. From (6.4), we get

$$
\begin{aligned}
P_Y(s) &= Es^Y = P_{X_1}(s)P_{X_2}(s) \\
&= \frac{1-p_1}{1-p_1 s} \times \frac{1-p_2}{1-p_2 s} \, .
\end{aligned}
\tag{6.23}
$$

The RHS of (6.23) can be simplified and written as

$$
P_Y(s) = \frac{p_1(1-p_2)}{p_1 - p_2} P_{X_1}(s) - \frac{p_2(1-p_1)}{p_1 - p_2} P_{X_2}(s).
\tag{6.24}
$$

Now it follows from (6.24) that

$$
\begin{aligned}
P\{Y = m\} &= \frac{p_1(1-p_2)}{p_1 - p_2} P\{X_1 = m\} - \frac{p_2(1-p_1)}{p_1 - p_2} P\{X_2 = m\} \\
&= \frac{p_1(1-p_1)(1-p_2)p_1^m}{p_1 - p_2} - \frac{p_2(1-p_2)(1-p_1)p_2^m}{p_1 - p_2} \\
&= \frac{(1-p_1)(1-p_2)}{p_1 - p_2} \left(p_1^{m+1} - p_2^{m+1} \right).
\end{aligned}
\tag{6.25}
$$

Indeed, (6.25) is also valid if $p_2 > p_1$. If $p_1 = p_2$, we must let $p_2 \to p_1$ in (6.25). As a result, the distribution of the sum of two independent $G(p)$ random variables is determined as

$$
P\{Y = m\} = (m+1)(1-p)^2 p^m, \qquad m = 0, 1, \ldots.
\tag{6.26}
$$

Below, we will obtain (6.26) as a special case of more general convolutions.

Now, let $X_1, X_2, \ldots, X_n$ be independent $G(p)$ random variables, and $Y_n = X_1 + \cdots + X_n$. The generating function of Y_n is given by

$$
P_n(s) = Es^{Y_n} = P_{X_1}(s) \cdots P_{X_n}(s) = \left(\frac{1-p}{1-ps} \right)^n.
\tag{6.27}
$$

Since

$$
(1-ps)^{-n} = \sum_{m=0}^{\infty} \binom{-n}{m} (-p)^m s^m,
\tag{6.28}
$$

we simply get

$$
P\{Y_n = m\} = (1-p)^n (-p)^m \binom{-n}{m},
\tag{6.29}
$$

where

$$
\begin{aligned}
\binom{-n}{m} &= \frac{(-n)(-n-1) \cdots (-n-m+1)}{m!} \\
&= (-1)^m \frac{n(n+1) \cdots (n+m-1)}{m!} \\
&= (-1)^m \binom{n+m-1}{n-1};
\end{aligned}
$$

hence,

$$P\{Y_n = m\} = \binom{n+m-1}{n-1} p^m (1-p)^n, \qquad m = 0, 1, \ldots. \qquad (6.30)$$

In particular, if $n = 2$, we arrive at (6.26).

In the scheme of independent trials, the random variable Y_n can be interpreted as the total number of "successes" until the appearance of nth "failure". Since the generating function of Y_n is a binomial of the negative order $(-n)$, the distribution in (6.30), called the *negative binomial distribution*, is discussed in detail in Chapter 7. The geometric distribution is naturally a particular case of the negative binomial distribution when $n = 1$.

6.7   Decompositions

It is easy to see that geometric $G(p)$ distribution can be represented as a sum of two or more independent random variables. In fact, the generating function $P_X(s) = (1-p)/(1-ps)$ of the random variable $X \sim G(p)$ can be rewritten as a product of two generating functions:

$$P_X(s) = P_U(s) P_V(s),$$

where

$$P_U(s) = \frac{1-p^2}{1-p^2 s^2} \qquad (6.31)$$

and

$$P_V(s) = \frac{1+ps}{1+p} . \qquad (6.32)$$

It is easy to see that the generating function in (6.31) corresponds to a random variable U, which has a geometric distribution, denoted by $G(p^2, 0, 2)$; i.e., U takes on values $0, 2, 4, \ldots$, with probabilities

$$r_m = P\{U = 2m\} = (1-p^2)p^{2m}, \qquad m = 0, 1, \ldots. \qquad (6.33)$$

Note that $U \overset{d}{=} 2W$, where $W \sim G(p^2)$. Further, the generating function $P_V(s)$ in (6.32) corresponds to Bernoulli $Be\left(\frac{p}{1+p}\right)$ distribution; i.e.,

$$P\{V = 1\} = 1 - P\{\nu = 0\} = \frac{p}{1+p} . \qquad (6.34)$$

Thus, for a $G(p)$ distributed random variable X, we get the representation

$$X \overset{d}{=} U + V,$$

where the distributions of the independent random variables U and V are as given in (6.33) and (6.34). We can also write it equivalently as

$$X \stackrel{d}{=} 2U_1 + V_1, \tag{6.35}$$

where $X \sim G(p)$, $U_1 \sim G(p^2)$ and $V_1 \sim Be\left(\dfrac{p}{1+p}\right)$. Moreover, U_1 in (6.35) is also a geometrically distributed random variable and itself admits the representation

$$U_1 \stackrel{d}{=} 2U_2 + V_2,$$

where $U_2 \sim G(p^4)$ and $V_2 \sim Be\left(\dfrac{p^2}{1+p^2}\right)$. Combining these, we have a new decomposition for X as

$$X \stackrel{d}{=} 4U_2 + 2V_2 + V_1.$$

This process can be continued and for any $n = 1, 2, \ldots$, the following decomposition holds:

$$X \stackrel{d}{=} 2^n U_n + \sum_{k=1}^{n} 2^{k-1} V_k, \tag{6.36}$$

where

$$U_n \sim G(p^{2^n}) \quad \text{and} \quad V_k \sim Be\left(\frac{p^{2^{k-1}}}{1 + p^{2^{k-1}}}\right).$$

Note that all the random variables on the RHS of (6.36) are independent.

In Chapter 7, when we introduce the general form of negative binomial distributions, it will be shown that any geometric distribution is infinitely divisible.

6.8   Entropy

For geometric distributions, one can easily derive the entropy.

Exercise 6.3 Show that the entropy of $X \sim G(p)$, $0 < p < 1$, is given by

$$H(X) = -\ln(1 - p) - \frac{p}{1 - p} \ln p; \tag{6.37}$$

in particular, $H(X) = 2 \ln 2$ if $p = \frac{1}{2}$.

6.9 Conditional Probabilities

In Chapter 5 we derived the hypergeometric distribution by considering the conditional distribution of X_1, given that $X_1 + X_2$ is fixed. We will now try to find the corresponding conditional distribution in the case when we have geometric distributions.

Let X_1 and X_2 be independent and identically distributed $G(p)$ random variables, and $Y = X_1 + X_2$. In our case, $Y = 0, 1, 2, \ldots$ and [see (6.30) when $n = 2$]

$$P\{Y = r\} = (r + 1)(1 - p)^2 p^r, \qquad r = 0, 1, 2, \ldots.$$

We can find the required conditional probabilities as

$$
\begin{aligned}
P\{X_1 = t | X_1 + X_2 = r\} &= \frac{P\{X_1 = t, \ X_2 = r - t\}}{P\{X_1 + X_2 = r\}} \\
&= \frac{(1 - p)p^t (1 - p)p^{r-t}}{(r + 1)(1 - p)^2 p^r} \\
&= \frac{1}{r + 1}.
\end{aligned}
\tag{6.38}
$$

It follows from (6.38) that the conditional distribution of X_1, given that $X_1 + X_2 = r$, becomes the discrete uniform $DU(r + 1)$ distribution.

Geometric distributions possess one more interesting property concerning conditional probabilities.

Exercise 6.4 Let $X \sim G(p)$, $0 < p < 1$. Show that

$$P\{X \geq n + m | X \geq m\} = P\{X \geq n\} \tag{6.39}$$

holds for any $n = 0, 1, \ldots$ and $m = 0, 1, \ldots$.

Remark 6.1 Imagine that in a sequence of independent trials, we have m "successes" in a row and no "failures". It turns out that the additional number of "successes" until the first "failure" that we shall observe now, has the same geometric $G(p)$ distribution. Among all discrete distributions, the geometric distribution is the only one which possesses this *lack of memory property* in (6.39).

Remark 6.2 Instead of defining a geometric random variable as the number of "successes" until the appearance of the first "failure" (denoted by X), we could define it alternatively as the number of "trials" required for the first "failure" (denoted by Z). It is clear that $Z = X + 1$ so that we readily have the mass function of Z as [see (6.1)]

$$P\{Z = n\} = (1 - p)p^{n-1}, \qquad n = 1, 2, \ldots. \tag{6.40}$$

From (6.4), we then have the generating function of Z as

$$P_Z(s) = Es^Z = Es^{X+1} = sEs^X = \frac{s(1-p)}{1-ps} \qquad \text{if } |s| < \frac{1}{p}. \qquad (6.41)$$

Many authors, in fact, take these as 'standard form' of the geometric distribution [instead of (6.1) and (6.4)].

6.10 Geometric Distribution of Order k

Consider a sequence of Bernoulli(p) trials. Let Z_k denote the number of "trials" required for "k consecutive failures" to appear for the first time. Then, it can be shown that the generating function of Z_k is given by

$$P_{Z_k}(s) = Es^{Z_k} = \frac{(1-p)^k s^k (1-s+ps)}{1-s+p(1-p)^k s^{k+1}}. \qquad (6.42)$$

The corresponding distribution has been named the *geometric distribution of order k*. Clearly, the generating function in (6.42) reduces to that of the geometric distribution in (6.41) when $k = 1$. For a review of this and many other "run-related" distributions, one may refer to the recent book by Balakrishnan and Koutras (2002).

Exercise 6.5 Show that (6.42) is indeed the generating function of Z_k.

Exercise 6.6 From the generating function of Z_k in (6.42), derive the mean and variance of Z_k.

CHAPTER 7

NEGATIVE BINOMIAL DISTRIBUTION

7.1 Introduction

In Chapter 6 we discussed the convolution of n geometric $G(p)$ distributions and derived a new distribution in (6.30) whose generating function has the form

$$P_n(s) = (1 - p)^n (1 - ps)^{-n}.$$

As we noted there, for any positive integer n, the corresponding random variable has a suitable interpretation in the scheme of independent trials as the distribution of the total number of "successes" until the nth "failure". For this interpretation, of course, n has to be an integer, but it will be interesting to see what will happen if we take the binomial of arbitrary negative order as the generating function of some random variable.

Specifically, let us consider the function

$$P_\alpha(s) = (1 - p)^\alpha (1 - ps)^{-\alpha}, \qquad \alpha > 0, \tag{7.1}$$

which is a generating function in the case when all the coefficients $p_0, p_1, \ldots$ in the power expansion $\sum_{k=0}^{\infty} p_k s^k$ of the RHS of (7.1) are nonnegative and their sum equals 1. The second condition holds since $\sum_{k=0}^{\infty} p_k = P_\alpha(1) = 1$ for any $\alpha > 0$. The function in (7.1) has derivatives of all orders, and simple calculations enable us to find the coefficients p_k in the expansion

$$P_\alpha(s) = \sum_{k=0}^{\infty} p_k s^k$$

as

$$
\begin{aligned}
p_k &= \frac{P_\alpha^{(k)}(0)}{k!} = \frac{(1 - p)^\alpha (-p)^k (-\alpha)(-\alpha - 1) \cdots (-\alpha - k + 1)}{k!} \\
&= \frac{\alpha(\alpha + 1) \cdots (\alpha + k - 1)}{k!} (1 - p)^\alpha p^k \\
&= \binom{\alpha + k - 1}{k} (1 - p)^\alpha p^k
\end{aligned}
\tag{7.2}
$$

and $p_k > 0$ for any $k = 0, 1, 2, \ldots$. Thus, we have obtained the following assertion: For any $\alpha > 0$, the function in (7.1) is generating the distribution concentrated at points $0, 1, 2, \ldots$ with probabilities as in (7.2).

It should be mentioned here that negative binomial distributions and some of their generalized forms have found important applications in actuarial analysis; see, for example, the book of Klugman, Panjer and Willmot (1998).

7.2   Notations

Let a random variable X take on values $0, 1, \ldots$ with probabilities

$$p_k = P\{X = k\} = \binom{\alpha + k - 1}{k}(1 - p)^\alpha p^k, \qquad \alpha > 0.$$

We say that X has the *negative binomial distribution* with parameters α and p ($\alpha > 0$, $0 < p < 1$), and denote it by

$$X \sim NB(\alpha, p).$$

Note that $NB(1, p) = G(p)$ and, hence, the geometric distribution is a particular case of the negative binomial distribution. Sometimes, the negative binomial distribution with an integer parameter α is called a *Pascal distribution*.

7.3   Generating Function and Characteristic Function

There is no necessity to calculate the generating function of this distribution. Unlike the earlier situations, the generating function was the primary object in this case, which then yielded the probabilities. Recall that if $X \sim NB(\alpha, p)$, then

$$P_X(s) = Es^X = \left(\frac{1 - p}{1 - ps}\right)^\alpha. \tag{7.3}$$

From (7.3), we immediately have the characteristic function of X as

$$f_X(t) = Ee^{itX} = P_X(e^{it}) = \left(\frac{1 - p}{1 - pe^{it}}\right)^\alpha. \tag{7.4}$$

7.4   Moments

From the expression of the generating function in (7.3), we can readily find the factorial moments.

Factorial moments of positive order:

$$\begin{aligned}
\mu_k &= EX(X-1)\cdots(X-k+1) = P_X^{(k)}(1)\\
&= \alpha(\alpha+1)\cdots(\alpha+k-1)\left(\frac{p}{1-p}\right)^k\\
&= \frac{\Gamma(\alpha+k)}{\Gamma(\alpha)} \times \frac{p^k}{(1-p)^k}, \quad k=1,2,\ldots,
\end{aligned} \tag{7.5}$$

where

$$\Gamma(s) = \int_0^\infty e^{-x} x^{s-1}\,dx$$

is the complete gamma function. In particular, we have

$$\mu_1 = EX = \frac{\alpha p}{1-p}, \tag{7.6}$$

$$\mu_2 = \frac{\alpha(\alpha+1)p^2}{(1-p)^2}, \tag{7.7}$$

$$\mu_3 = \frac{\alpha(\alpha+1)(\alpha+2)p^3}{(1-p)^3}, \tag{7.8}$$

$$\mu_4 = \frac{\alpha(\alpha+1)(\alpha+2)(\alpha+3)p^4}{(1-p)^4}. \tag{7.9}$$

Note that if $\alpha = 1$, then (7.5)–(7.9) reduce to the corresponding moments for the geometric distribution given in (6.6)–(6.10), respectively.

Factorial moments of negative order:

$$\mu_{-k} = \int_0^1 \int_0^{t_{k-1}} \cdots \int_0^{t_1} \left(\frac{1-p}{1-ps}\right)^\alpha ds\,dt_1\cdots dt_{k-1}, \tag{7.10}$$

and in particular,

$$\begin{aligned}
\mu_{-1} &= E\left(\frac{1}{X+1}\right) = (1-p)^\alpha \int_0^1 (1-ps)^{-\alpha}\,ds\\
&= \frac{(1-p)^\alpha}{p(1-\alpha)}\left\{1-(1-p)^{1-\alpha}\right\}
\end{aligned} \tag{7.11}$$

if $\alpha > 0$ and $\alpha \ne 1$. The expression for μ_{-1} for the case $\alpha = 1$ is as given in (6.12).

Moments about zero:

$$\alpha_1 = EX = \mu_1 = \frac{\alpha p}{1-p}, \tag{7.12}$$

$$\alpha_2 = EX^2 = \mu_1 + \mu_2 = \frac{\alpha p(1+\alpha p)}{(1-p)^2}, \tag{7.13}$$

$$\begin{aligned}
\alpha_3 &= EX^3 = \mu_1 + 3\mu_2 + \mu_3\\
&= \frac{\alpha p}{(1-p)^3}\left\{1+(3\alpha+1)p+\alpha^2 p^2\right\},
\end{aligned} \tag{7.14}$$

$$\begin{aligned}
\alpha_4 &= EX^4 = \mu_1 + 7\mu_2 + 6\mu_3 + \mu_4\\
&= \frac{\alpha p}{(1-p)^4}\left\{1+(7\alpha+4)p+(6\alpha^2+4\alpha+1)p^2+\alpha^3 p^3\right\}.
\end{aligned} \tag{7.15}$$

Central moments:

From (7.12) and (7.13), we readily find the variance of ξ as

$$\beta_2 = \text{Var } X = \alpha_2 - \alpha_1^2 = \frac{\alpha p}{(1-p)^2} \,. \qquad (7.16)$$

Similarly, from (7.12)–(7.15), we find the third and fourth central moments of X as

$$\beta_3 \;=\; \alpha_3 - 3\alpha_2\alpha_1 + 2\alpha_1^3 = \frac{\alpha p}{(1-p)^3}(1+p), \qquad (7.17)$$

$$\beta_4 \;=\; \alpha_4 - 4\alpha_3\alpha_1 + 6\alpha_2\alpha_1^2 - 3\alpha_1^4 = \frac{\alpha p}{(1-p)^4}\left\{1 + (3\alpha + 4)p + p^2\right\}, \qquad (7.18)$$

respectively.

Shape characteristics:

From (7.16)–(7.18), we find Pearson's coefficients of skewness and kurtosis as

$$\gamma_1 = \frac{\beta_3}{\beta_2^{3/2}} = \frac{1+p}{\sqrt{\alpha p}} \qquad (7.19)$$

and

$$\begin{aligned}
\gamma_2 = \frac{\beta_4}{\beta_2^2} &= \frac{1 + (3\alpha + 4)p + p^2}{\alpha p} \\
&= 3 + \left(\frac{1 + 4p + p^2}{\alpha p}\right), \qquad (7.20)
\end{aligned}$$

respectively. It is quite clear from (7.19) that the negative binomial distribution is positively skewed. Furthermore, we observe from (7.20) that the distribution is leptokurtic for all values of the parameters α and p. We also observe that as α tends to ∞, γ_1 and γ_2 in (7.19) and (7.20) tend to 0 and 3 (the values corresponding to the normal distribution), respectively. Plots of negative binomial mass function presented in Figures 7.1–7.3 reveal these properties.

7.5   Convolutions and Decompositions

Let $X_1 \sim NB(\alpha_1, p)$ and $X_2 \sim NB(\alpha_2, p)$ be two independent random variables, and $Y = X_1 + X_2$.

Exercise 7.1 Use (7.3) to establish that Y has a negative binomial $NB(\alpha_1 + \alpha_2, p)$ distribution.

NegBin(2,0.25)

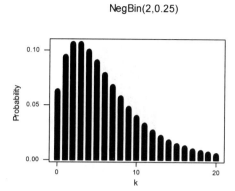

NegBin(2,0.5)

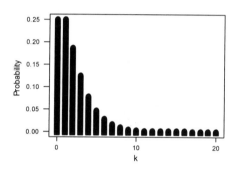

NegBin(2,0.75)

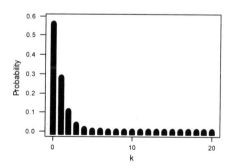

Figure 7.1. Plots of negative binomial mass function when $r = 2$

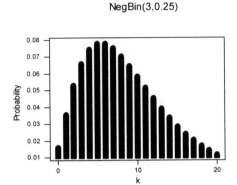

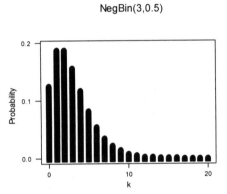

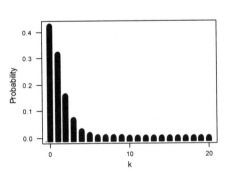

Figure 7.2. Plots of negative binomial mass function when $r = 3$

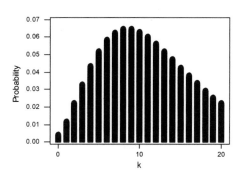

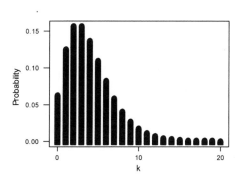

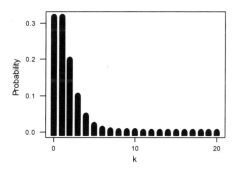

Figure 7.3. Plots of negative binomial mass function when $r = 4$

Remark 7.1 Now we see that the sum of two or more independent random variables $X_k \sim NB(\alpha_k, p)$, $k = 1, 2, \ldots$, also has a negative binomial distribution. On the other hand, the assertion above enables us to conclude that any negative binomial distribution admits a decomposition with negative binomial components. In fact, for any $n = 1, 2, \ldots$, a random variable $X \sim N(\alpha, p)$ can be represented as

$$X \stackrel{d}{=} X_{1,n} + \cdots + X_{n,n},$$

where $X_{1,n}, \ldots, X_{n,n}$, are independent and identically distributed random variables having $NB(\alpha/n, p)$ distribution. This means that any negative binomial distribution (including geometric) is infinitely divisible.

7.6   Tail Probabilities

Let $X \sim NB(n, p)$, where n is a positive integer. The interpretation of X based on independent trials (each resulting in "successes" and "failures") gives us a way to obtain a simple form for tail probabilities as

$$P\{X \geq m\} = \sum_{k=m}^{\infty} \binom{n+k-1}{k} (1-p)^n p^k. \qquad (7.21)$$

In fact, event $\{X \geq m\}$ is equivalent to the following: If we fix outcomes of the first $m + n - 1$ trials, then the number of "successful" trials must be at least m. Let Y be the number of "successes" in $m + n - 1$ independent trials. We know that Y has the binomial $B(m + n - 1, p)$ distribution and

$$P\{Y \geq m\} = \sum_{k=m}^{m+n-1} \binom{m+n-1}{k} p^k (1-p)^{m+n-1-k}. \qquad (7.22)$$

Moreover, (5.36) gives the following expression for the RHS of (7.22):

$$\sum_{k=m}^{m+n-1} \binom{m+n-1}{k} p^k (1-p)^{m+n-1-k}$$
$$= \frac{(m+n-1)!}{(m-1)!\,(n-1)!} \int_0^p x^{m-1} (1-x)^{n-1}\, dx. \qquad (7.23)$$

Since

$$P\{X \geq m\} = P\{Y \geq m\},$$

upon combining (7.21)–(7.23), we obtain

$$P\{X \geq m\} = \sum_{k=m}^{\infty} \binom{n+k-1}{k} (1-p)^n p^k$$
$$= \frac{(m+n-1)!}{(m-1)!\,(n-1)!} \int_0^p x^{m-1} (1-x)^{n-1}\, dx. \qquad (7.24)$$

As a special case of (7.24), we get equality (6.3) for geometric $G(p)$ distribution when $n = 1$.

7.7 Limiting Distributions

Fox $\lambda > 0$, let us consider $X_\alpha \sim NB(\alpha, \lambda/\alpha)$ where $\lambda/\alpha < 1$. The generating function of X_α in this case is

$$P_\alpha(s) = Es^{X_\alpha} = \left(1 - \frac{\lambda}{\alpha}\right)^\alpha \bigg/ \left(1 - \frac{\lambda s}{\alpha}\right)^\alpha. \tag{7.25}$$

We see immediately that

$$P_\alpha(s) \to e^{\lambda(s-1)} \quad \text{as } \alpha \to \infty. \tag{7.26}$$

Note that

$$e^{\lambda(s-1)} = \sum_{k=0}^\infty \frac{e^{-\lambda}\lambda^k}{k!} s^k \tag{7.27}$$

is the generating function of a random variable Y taking on values $0, 1, 2, \ldots$ with probabilities

$$p_k = \frac{e^{-\lambda}\lambda^k}{k!}, \qquad k = 0, 1, 2, \ldots. \tag{7.28}$$

In Chapter 5 we mentioned that Y has the *Poisson distribution*. Now relation (7.26) implies that for any $k = 0, 1, \ldots,$

$$P\{X_\alpha = k\} \to \frac{e^{-\lambda}\lambda^k}{k!} \quad \text{as } \alpha \to \infty; \tag{7.29}$$

that is, the Poisson distribution is the limit for the sequence of $NB(\alpha, \lambda/\alpha)$ distributions as $\alpha \to \infty$.

Next, let $X_\alpha \sim NB(\alpha, p)$ and

$$\begin{aligned} W_\alpha &= \frac{X_\alpha - EX_\alpha}{\sqrt{\operatorname{Var} X_\alpha}} \\ &= \left(X_\alpha - \frac{\alpha p}{1-p}\right)\frac{1-p}{\sqrt{\alpha p}}. \end{aligned} \tag{7.30}$$

Let $f_\alpha(t)$ be the characteristic function of W_α, which can be derived by standard methods from (7.4). It turns out that for any t,

$$f_\alpha(t) \to \exp\left\{-\frac{t^2}{2}\right\} \quad \text{as } \alpha \to \infty. \tag{7.31}$$

Comparing this with (5.43), we note that the sequence W_α, as $\alpha \to \infty$, converges in distribution to the standard normal distribution, which we came across in Chapter 5 in a similar context.

CHAPTER 8

HYPERGEOMETRIC DISTRIBUTION

8.1   Introduction

In Chapter 5 we derived hypergeometric distribution as the conditional distribution of X_1, given that $X_1 + X_2$ is fixed, where X_1 and X_2 were binomial random variables. A simpler situation wherein hypergeometric distributions arise is in connection with classical combinatorial problems. Suppose that an urn contains a red and b black balls. Suppose that n balls are drawn at random from the urn (without replacement). Let X be the number of red balls in the sample drawn. Then, it is clear that X takes on an integer value m such that

$$\max(0, n - b) \le m \le \min(n, a), \tag{8.1}$$

with probabilities

$$p_m = P\{X = m\} = \binom{a}{m}\binom{b}{n - m} \bigg/ \binom{a + b}{n}. \tag{8.2}$$

8.2   Notations

In this case we say that X has a *hypergeometric distribution* with parameters n, a, and b, and denote it by

$$X \sim Hg(n, a, b).$$

Remark 8.1 Inequalities (8.1) give those integers m for which the RHS of (8.2) is nonzero. We have from (8.2) the following identity:

$$\sum_{m=\max(0, n-b)}^{\min(n, a)} \binom{a}{m}\binom{b}{n - m} = \binom{a + b}{n}. \tag{8.3}$$

To simplify our calculations, we will suppose in the sequel that $n \leq \min(a, b)$ and hence probabilities in (8.2) are positive for $m = 0, 1, \ldots, n$ only. Identity (8.3) then becomes the following useful equality:

$$\sum_{m=0}^{n} \binom{a}{m} \binom{b}{n-m} = \binom{a+b}{n}. \tag{8.4}$$

8.3   Generating Function

If $X \sim Hg(n, a, b)$, then we can write a formal expression for its generating function as

$$P_X(s) = \frac{n! \, (a+b-n)!}{(a+b)!} \sum_{m=0}^{n} \frac{a! \, b!}{m! \, (a-m)! \, (n-m)! \, (b-n+m)!} s^m. \tag{8.5}$$

It turns out that the RHS of (8.5) can be simplified if we use the Gaussian hypergeometric function

$$_2F_1(\alpha, \beta; \gamma; s) = 1 + \frac{\alpha\beta}{\gamma} \frac{s}{1!} + \frac{\alpha(\alpha+1)\beta(\beta+1)}{\gamma(\gamma+1)} \frac{s^2}{2!} + \cdots \,,$$

which was introduced in Chapter 2. Then, the generating function in (8.5) becomes

$$P_X(s) = \frac{_2F_1[-n, -a; b-n+1; s]}{_2F_1[-n, -a; b-n+1; 1]}, \tag{8.6}$$

from which it becomes clear why the distribution has been given the name *hypergeometric distribution*.

8.4   Characteristic Function

On applying the relation between generating function and characteristic function, we readily obtain the characteristic function from (8.6) to be

$$\begin{aligned} f_X(t) &= E e^{itX} = P_X(e^{it}) \\ &= \frac{_2F_1[-n, -a; b-n+1; e^{it}]}{_2F_1[-n, -a; b-n+1; 1]}. \end{aligned} \tag{8.7}$$

8.5   Moments

To begin with, we show how we can obtain the most important moments $\alpha_1 = EX$ and $\beta_2 = \text{Var } X$ using the "urn interpretation" of X. We have a red balls in the urn, numbered $1, 2, \ldots, a$. Corresponding to each of the red balls, we introduce the random indicators $Y_1, Y_2, \ldots, Y_a$ as follows:

$$
\begin{aligned}
Y_k \;&=\; 1 \quad \text{if the } k\text{th red ball is drawn} \\
&=\; 0 \quad \text{otherwise.}
\end{aligned}
$$

Note that

$$
EY_k = P\{Y_k = 1\} = \frac{n}{a+b}, \qquad k = 1, 2, \ldots, a, \tag{8.8}
$$

and

$$
\text{Var } Y_k = \frac{n}{a+b}\left(1 - \frac{n}{a+b}\right), \qquad k = 1, 2, \ldots, a. \tag{8.9}
$$

It is not difficult to see that

$$
X = Y_1 + Y_2 + \cdots + Y_a. \tag{8.10}
$$

It follows immediately from (8.10) that

$$
EX = E(Y_1 + \cdots + Y_a) = a EY_1 = \frac{an}{a+b}. \tag{8.11}
$$

Now,

$$
\text{Var } X = \sum_{k=1}^{a} \text{Var } Y_k + 2 \sum_{1 \le i < k \le a} \text{Cov}(Y_i, Y_k). \tag{8.12}
$$

Using the symmetry argument, we can rewrite (8.12) as

$$
\text{Var } X = a \text{ Var } Y_1 + a(a-1)\,\text{Cov}(Y_1, Y_2). \tag{8.13}
$$

Now we only need to find the covariance on the RHS of (8.13). For this purpose, we have

$$
\begin{aligned}
\text{Cov}(Y_1, Y_2) \;&=\; E(Y_1 Y_2) - EY_1 EY_2 \\
&=\; P\{Y_1 Y_2 = 1\} - \left(\frac{n}{a+b}\right)^2.
\end{aligned}
\tag{8.14}
$$

Note that

$$
P\{Y_1 Y_2 = 1\} = \binom{n}{2} \Big/ \binom{a+b}{2} = \frac{n(n-1)}{(a+b)(a+b-1)}. \tag{8.15}
$$

Finally, (8.9) and (8.13)–(8.15) readily yield

$$
\text{Var } X = \frac{abn(a+b-n)}{(a+b)^2(a+b-1)}. \tag{8.16}
$$

Factorial moments of positive order:

$$\mu_k = EX(X-1)\cdots(X-k+1)$$

$$= \sum_{m=0}^{n} m(m-1)\cdots(m-k+1)\binom{a}{m}\binom{b}{n-m}\Bigg/\binom{a+b}{n}$$

$$= \sum_{m=k}^{n} m(m-1)\cdots(m-k+1)\binom{a}{m}\binom{b}{n-m}\Bigg/\binom{a+b}{n}$$

$$= \frac{a!\,b!\,n!\,(a+b-n)!}{(a+b)!} \sum_{m=k}^{n} \frac{1}{(m-k)!\,(a-m)!\,(n-m)!\,(b-n+m)!}$$

$$= \frac{a!\,b!\,n!\,(a+b-n)!}{(a+b)!} \sum_{m=0}^{n-k} \frac{1}{m!\,(a-k-m)!\,(n-k-m)!\,(b-n+k+m)!}$$

$$= \frac{a!\,n!\,(a+b-n)!}{(a+b)!\,(a-k)!} \sum_{m=0}^{n-k} \binom{a-k}{m}\binom{b}{n-k-m}. \tag{8.17}$$

From (8.4), we know that

$$\sum_{m=0}^{n-k} \binom{a-k}{m}\binom{b}{n-k-m} = \binom{a+b-k}{n-k} = \frac{(a+b-k)!}{(n-k)!\,(a+b-n)!}$$

using which we readily obtain

$$\mu_k = \frac{a!\,n!\,(a+b-k)!}{(a+b)!\,(a-k)!\,(n-k)!} \qquad \text{for } k \le n. \tag{8.18}$$

Note also that $\mu_k = 0$ if $k > n$.

In particular, the first four factorial moments are as follows:

$$\mu_1 = \frac{an}{a+b}, \tag{8.19}$$

$$\mu_2 = \frac{a(a-1)n(n-1)}{(a+b)(a+b-1)}, \tag{8.20}$$

$$\mu_3 = \frac{a(a-1)(a-2)n(n-1)(n-2)}{(a+b)(a+b-1)(a+b-2)}, \tag{8.21}$$

$$\mu_4 = \frac{a(a-1)(a-2)(a-3)n(n-1)(n-2)(n-3)}{(a+b)(a+b-1)(a+b-2)(a+b-3)}. \tag{8.22}$$

Factorial moments of negative order:

Analogous to (8.17) and (8.18), we have

$$\mu_{-k} = E\left[\frac{1}{(X+1)(X+2)\cdots(X+k)}\right]$$

$$= \sum_{m=0}^{n} \binom{a}{m}\binom{b}{n-m}\Bigg/\left\{(m+1)\cdots(m+k)\binom{a+b}{n}\right\}$$

$$= \frac{a!\, n!\, (a+b-n)!}{(a+k)!\, (a+b)!} \sum_{m=k}^{n+k} \binom{a+k}{m} \binom{b}{n+k-m}$$

$$= \frac{a!\, n!\, (a+b-n)!}{(a+k)!\, (a+b)!} \binom{a+b+k}{n+k}$$

$$- \frac{a!\, n!\, (a+b-n)!}{(a+k)!\, (a+b)!} \sum_{m=0}^{k-1} \binom{a+k}{m} \binom{b}{n+k-m}. \tag{8.23}$$

In particular, we find

$$\mu_{-1} = E\left(\frac{1}{X+1}\right)$$

$$= \frac{a+b+1}{(a+1)(n+1)} - \frac{a!\, n!\, (a+b-n)!\, b!}{(a+1)!\, (a+b)!\, (n+1)!\, (b-n-1)!}$$

$$= \frac{a+b+1}{(a+1)(n+1)} - \frac{b!\, (a+b-n)!}{(a+1)(n+1)(a+b)!\, (b-n-1)!}. \tag{8.24}$$

Moments about zero:
  Indeed, the RHSs of (8.11) and (8.19) coincide, and we have

$$\alpha_1 = \mu_1 = \frac{an}{a+b}.$$

Furthermore, we can show that

$$\alpha_2 = EX^2 = \mu_1 + \mu_2 = \frac{an}{a+b} + \frac{a(a-1)n(n-1)}{(a+b)(a+b-1)} \tag{8.25}$$

and

$$\alpha_3 = EX^3 = \mu_1 + 3\mu_2 + \mu_3$$
$$= \frac{an}{a+b} + \frac{3a(a-1)n(n-1)}{(a+b)(a+b-1)} + \frac{a(a-1)(a-2)n(n-1)(n-2)}{(a+b)(a+b-1)(a+b-2)}. \tag{8.26}$$

Remark 8.2 If $n = 1$, then the hypergeometric $Hg(1, a, b)$ distribution coincides with Bernoulli $Be\left(\dfrac{a}{a+b}\right)$ distribution.

Exercise 8.1 Check that expressions for moments are the same for $Hg(1, a, b)$ as those for $Be\left(\dfrac{a}{a+b}\right)$ distributions.

Exercise 8.2 Derive the expressions of the second and third moments in (8.25) and (8.26), respectively.

8.6   Limiting Distributions

Let $X_N \sim Hg(n, pN, (1-p)N)$, $N = 1, 2, \ldots$, where $0 < p < 1$ and n is fixed. Then, we have

$$P\{X_N = m\}$$

$$= \binom{pN}{m}\binom{(1-p)N}{n-m} \bigg/ \binom{N}{n}$$

$$= \frac{n!}{m!\,(n-m)!}$$

$$\times \frac{pN(pN-1)\cdots(pN-m+1)((1-p)N)((1-p)N-1)\cdots((1-p)N-n+m+1)}{N(N-1)\cdots(N-n+1)},$$

from which it is easy to see that for any fixed $m = 0, 1, \ldots, n$,

$$P\{X_N = m\} \to \binom{n}{m}p^m(1-p)^{n-m} \qquad \text{as } N \to \infty. \tag{8.27}$$

Thus, we get the binomial distribution as a limit of a special sequence of hypergeometric distributions.

Exercise 8.3 Let $X_N \sim Hg(N, \lambda N^2, N^3)$, $N = 1, 2, \ldots$. Show then that for any fixed $m = 0, 1, \ldots$,

$$P\{X_N = m\} \to \frac{\lambda^m}{m!}\,e^{-\lambda} \qquad \text{as } N \to \infty. \tag{8.28}$$

The Poisson distribution, which is the limit for the sequence of hypergeometric random variables, present in Exercise 8.3, is the subject of discussion in the next chapter.

CHAPTER 9

POISSON DISTRIBUTION

9.1 Introduction

The Poisson distribution arises naturally in many instances; for example, as we have already seen in preceding chapters, it appears as a limiting distribution of some sequences of binomial, negative binomial, and hypergeometric random variables. In addition, due to its many interesting characteristic properties, it is also used as a probability model for the occurrence of rare events.

A book-length account of Poisson distributions, discussing in great detail their various properties and applications, is available [Haight (1967)].

9.2 Notations

Let a random variable X take on values $0, 1, \ldots$ with probabilities

$$p_m = P\{X = m\} = \frac{e^{-\lambda}\lambda^m}{m!}, \qquad m = 0, 1, \ldots, \tag{9.1}$$

where $\lambda > 0$. We say that X has a *Poisson distribution* with parameter λ, and denote it by

$$X \sim \pi(\lambda).$$

If $Y = a + hX$, $-\infty < a < \infty$, $h > 0$, then Y takes on values $a, a+h, a+2h, \ldots$ with probabilities

$$P\{Y = a + mh\} = P\{X = m\} = \frac{e^{-\lambda}\lambda^m}{m!}, \qquad m = 0, 1, \ldots.$$

This distribution also belongs to the Poisson type of distribution, and it will be denoted by $\pi(\lambda, a, h)$. The standard Poisson $\pi(\lambda)$ distribution is nothing but $\pi(\lambda, 0, 1)$.

9.3 Generating Function and Characteristic Function

Let $X \sim \pi(\lambda)$, $\lambda > 0$. From (9.1), we obtain the generating function of X as

$$P_X(s) = Es^X = \sum_{m=0}^{\infty} e^{-\lambda} \frac{\lambda^m}{m!} s^m$$

$$= e^{-\lambda} \sum_{m=0}^{\infty} \frac{(\lambda s)^m}{m!} = e^{\lambda(s-1)}. \tag{9.2}$$

Then, the characteristic function of X has the form

$$f_X(t) = Ee^{itX} = P_X(e^{it}) = \exp\{\lambda(e^{it} - 1)\}. \tag{9.3}$$

9.4 Moments

The simple form of the generating function in (9.2) enables us to derive all the factorial moments easily.

Factorial moments of positive order:

$$\mu_k = EX(X-1)\cdots(X-k+1) = P_X^{(k)}(1) = \lambda^k, \quad k = 1, 2, \ldots. \tag{9.4}$$

In particular, we have:

$$\mu_1 = \lambda, \tag{9.5}$$
$$\mu_2 = \lambda^2, \tag{9.6}$$
$$\mu_3 = \lambda^3, \tag{9.7}$$
$$\mu_4 = \lambda^4. \tag{9.8}$$

Factorial moments of negative order:

$$\mu_{-k} = E\left[\frac{1}{(X+1)(X+2)\cdots(X+k)}\right]$$

$$= \int_0^1 \int_0^{t_{k-1}} \cdots \int_0^{t_1} e^{\lambda(s-1)} \, ds \, dt_1 \cdots dt_{k-1}. \tag{9.9}$$

In particular, we obtain

$$\mu_{-1} = E\left[\frac{1}{X+1}\right] = \int_0^1 e^{\lambda(s-1)} \, ds = \frac{1 - e^{-\lambda}}{\lambda} \tag{9.10}$$

and

$$\mu_{-2} = E\left[\frac{1}{(X+1)(X+2)}\right] = \int_0^1 \int_0^t e^{\lambda(s-1)} \, ds \, dt$$

$$= \frac{1 - e^{-\lambda} - \lambda e^{-\lambda}}{\lambda^2}. \tag{9.11}$$

Moments about zero:

From (9.5)–(9.8), we immediately obtain the first four moments about zero as follows:

$$\alpha_1 = EX = \mu_1 = \lambda, \tag{9.12}$$

$$\alpha_2 = EX^2 = \mu_2 + \mu_1 = \lambda(\lambda + 1), \tag{9.13}$$

$$\alpha_3 = EX^3 = \mu_3 + 3\mu_2 + \mu_1 = \lambda(\lambda^2 + 3\lambda + 1), \tag{9.14}$$

$$\alpha_4 = EX^4 = \mu_4 + 6\mu_3 + 7\mu_2 + \mu_1 = \lambda(\lambda^3 + 6\lambda^2 + 7\lambda + 1). \tag{9.15}$$

Central moments:

From (9.12) and (9.13), we readily find the variance of X as

$$\beta_2 = \text{Var } X = \alpha_2 - \alpha_1^2 = \lambda. \tag{9.16}$$

Note that if $X \sim \pi(\lambda)$, then $EX = \text{Var } X = \lambda$.

Further, from (9.12)–(9.15), we also find the third and fourth central moments as

$$\beta_3 = E(X - EX)^3 = \alpha_3 - 3\alpha_2\alpha_1 + 2\alpha_1^3 = \lambda, \tag{9.17}$$

$$\beta_4 = E(X - EX)^4 = \alpha_4 - 4\alpha_3\alpha_1 + 6\alpha_2\alpha_1^2 - 3\alpha_1^4 = \lambda + 3\lambda^2, \tag{9.18}$$

respectively.

Shape characteristics:

From (9.16)–(9.18), we obtain the coefficients of skewness and kurtosis as

$$\gamma_1 = \frac{\beta_3}{\beta_2^{3/2}} = \frac{1}{\sqrt{\lambda}}, \tag{9.19}$$

$$\gamma_2 = \frac{\beta_4}{\beta_2^2} = 3 + \frac{1}{\lambda}, \tag{9.20}$$

respectively. From (9.19), we see that the Poisson distribution is positively skewed for all values of λ. Similarly, we see from (9.20) that the distribution is also leptokurtic for all values of λ. Furthermore, we observe that as λ tends to ∞, the values of γ_1 and γ_2 tend to 0 and 3 (the values corresponding to the normal distribution), respectively. Plots of Poisson mass function presented in Figure 9.1 reveal these properties.

9.5 Tail Probabilities

Let $X \sim \pi(\lambda)$. Then

$$P\{X \geq m\} = \sum_{r=m}^{\infty} \frac{e^{-\lambda}\lambda^r}{r!}, \qquad m = 0, 1, \ldots. \tag{9.21}$$

The RHS of (9.21) can be simplified and written in the form of an integral.

Exercise 9.1 Show that for any $m = 1, 2, \ldots$,

$$P\{X \geq m\} = \int_0^\lambda \frac{e^{-u} u^{m-1}}{(m-1)!}\, du. \tag{9.22}$$

Remark 9.1 We will recall the expression on the RHS of (9.22) later when we discuss the gamma distribution in Chapter 20.

9.6 Convolutions

Let $X_1 \sim \pi(\lambda_1)$ and $X_2 \sim \pi(\lambda_2)$ be independent random variables. Then, by making use of the generating functions $P_{X_1}(s)$ and $P_{X_2}(s)$, it is easy to show that $Y = X_1 + X_2$ has the generating function

$$P_Y(s) = P_{X_1}(s) \cdot P_{X_2}(s) = e^{(\lambda_1 + \lambda_2)(s-1)}$$

and, hence, $Y \sim \pi(\lambda_1 + \lambda_2)$. This simply means that convolutions of Poisson distributions are also distributed as Poisson.

9.7 Decompositions

Due to the result just stated, we know that for any $X \sim \pi(\lambda)$ we can find a pair of independent Poisson random variables $U \sim \pi(\lambda_1)$ and $V \sim \pi(\lambda - \lambda_1)$, where $0 < \lambda_1 < \lambda$, to obtain the decomposition

$$X \overset{d}{=} U + V. \tag{9.23}$$

Hence, any Poisson distribution is decomposable. Moreover, for any $n = 1, 2, \ldots$, the decomposition

$$X \overset{d}{=} X_1 + X_2 + \cdots + X_n \tag{9.24}$$

holds with X's being independent and identically distributed as $\pi(\lambda/n)$. This simply implies that X is infinitely divisible for any λ.

Raikov (1937b) established that if X admits decomposition (9.23), then both the independent nondegenerate components U and V have necessarily a Poisson type of distribution; that is, there exist constants $-\infty < a < \infty$ and $\lambda_1 < \lambda$ such that

$$U \sim \pi(\lambda_1, a, 1) \quad \text{and} \quad V \sim \pi(\lambda - \lambda_1, -a, 1).$$

Thus, convolutions and decompositions of Poisson distributions always belong to the Poisson class of distributions.

Poisson(1)

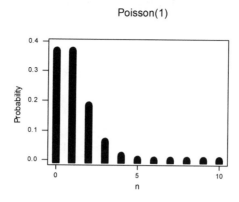

Poisson(4)

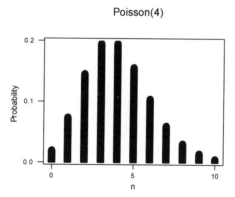

Poisson(10)

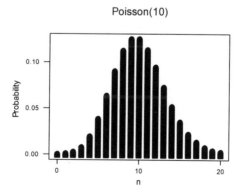

Figure 9.1. Plots of Poisson mass function

9.8   Conditional Probabilities

Let $X_1 \sim \pi(\lambda_1)$ and $X_2 \sim \pi(\lambda_2)$ be independent random variables. Consider the conditional distribution of X_1 given that $X_1 + X_2$ is fixed. Since $X_1 + X_2 \sim \pi(\lambda_1 + \lambda_2)$, we obtain for any $n = 0, 1, 2, \ldots$ and $m = 0, 1, \ldots, n$ that

$$
\begin{aligned}
& P\{X_1 = m | X_1 + X_2 = n\} \\
={} & \frac{P\{X_1 = m, X_1 + X_2 = n\}}{P\{X_1 + X_2 = n\}} \\
={} & \frac{P\{X_1 = m, X_2 = n - m\}}{P\{X_1 + X_2 = n\}} \\
={} & \frac{P\{X_1 = m\} P\{X_2 = n - m\}}{P\{X_1 + X_2 = n\}} \\
={} & \frac{e^{-\lambda_1} \lambda_1^m}{m!} \frac{e^{-\lambda_2} \lambda_2^{n-m}}{(n-m)!} \frac{n!}{e^{-(\lambda_1 + \lambda_2)}(\lambda_1 + \lambda_2)^n} \\
={} & \binom{n}{m} \left(\frac{\lambda_1}{\lambda_1 + \lambda_2} \right)^m \left(\frac{\lambda_2}{\lambda_1 + \lambda_2} \right)^{n-m}.
\end{aligned}
\tag{9.25}
$$

Thus, the conditional distribution of X_1, given that $X_1 + X_2 = n$, is simply the binomial $B\left(n, \dfrac{\lambda_1}{\lambda_1 + \lambda_2} \right)$ distribution.

Now, we will try to solve the inverse problem. Let X_1 and X_2 be independent random variables taking on values $0, 1, 2, \ldots$ with positive probabilities. Then, $Y = X_1 + X_2$ also takes on values $0, 1, 2, \ldots$ with positive probabilities. Suppose that for any $n = 0, 1, \ldots$, the conditional distribution of X_1, given that $Y = n$, is binomial $B(n, p)$ for some parameter p, $0 < p < 1$. Then, we are interested in determining the distributions of X_1 and X_2! It turns out that both distributions are Poisson.

To see this, let

$$
P\{X_1 = m\} = r_m > 0, \quad m = 0, 1, \ldots,
$$

and

$$
P\{X_2 = \ell\} = q_\ell > 0, \quad \ell = 0, 1, \ldots.
$$

As seen above, the conditional probabilities $P\{X_1 = m | Y = n\}$ result in the expression $r_m q_{n-m} / P\{Y = n\}$. In this situation, we get the equality

$$
\frac{r_m q_{n-m}}{P\{Y = n\}} = \frac{n!}{m! \, (n-m)!} p^m (1-p)^{n-m}, \quad m = 0, 1, \ldots, n.
\tag{9.26}
$$

Compare (9.26) with the same equality written for $m + 1$ in place of m, which has the form

$$
\frac{r_{m+1} q_{n-m-1}}{P\{Y = n\}} = \frac{n!}{(m+1)! \, (n-m-1)!} p^{m+1} (1-p)^{n-m-1}.
\tag{9.27}
$$

It readily follows from (9.26) and (9.27) that

$$
\frac{(n-m)p}{(m+1)(1-p)} = \frac{r_{m+1}}{r_m} \frac{q_{n-m-1}}{q_{n-m}}
\tag{9.28}
$$

holds for any $n = 1, 2, \ldots$ and $m = 0, 1, \ldots, n$. In particular, we can take $m = 0$ to obtain

$$q_n = \frac{r_1(1-p)}{r_0 n p} q_{n-1}, \qquad n = 1, 2, \ldots. \tag{9.29}$$

Let us denote $\lambda = r_1(1-p)/(r_0 p)$. Then, (9.29) implies that

$$q_n = \frac{\lambda}{n} q_{n-1} = \frac{\lambda^2}{n(n-1)} q_{n-2} = \cdots = \frac{\lambda^n}{n!} q_0, \qquad n = 1, 2, \ldots. \tag{9.30}$$

Since

$$1 = \sum_{n=0}^{\infty} q_n = q_0 \left(1 + \lambda + \frac{\lambda^2}{2!} + \cdots \right) = q_0 \, e^\lambda,$$

we immediately get $q_0 = e^{-\lambda}$ and, consequently,

$$q_n = \frac{e^{-\lambda} \lambda^n}{n!}, \qquad n = 0, 1, \ldots, \tag{9.31}$$

where λ is some positive constant. On substituting q_n into (9.28) and taking $m = n - 1$, we obtain

$$r_n = \frac{\lambda p}{n(1-p)} r_{n-1}, \qquad n = 1, 2, \ldots.$$

Hence,

$$r_n = \frac{[\lambda p/(1-p)]^n}{n!} r_0$$

and consequently,

$$r_n = \frac{e^{-\lambda}[\lambda p/(1-p)]^n}{n!}, \qquad n = 0, 1, 2, \ldots. \tag{9.32}$$

Thus, only for X_1 and X_2 having Poisson distributions, the conditional distribution of X_1, given that $X_1 + X_2$ is fixed, can be binomial.

9.9 Maximal Probability

We may often be interested to know what of probabilities $p_m = e^{-\lambda} \lambda^m / m!$ $(m = 0, 1, \ldots)$ are maximal.

Exercise 9.2 Show that there are the following two situations for maximal Poisson probabilities: If $m_0 < \lambda < m_0 + 1$, where m_0 is an integer, then p_{m_0} is maximal among all probabilities p_m $(m = 0, 1, \ldots)$. If $\lambda = m_0$, then $p_{m_0} = p_{m_0-1}$, and in this case both these probabilities are maximal.

9.10 Limiting Distribution

Let $X_n \sim \pi(n)$, $n = 1, 2, \ldots$, and

$$W_n = \frac{X_n - EX_n}{\sqrt{\text{Var } X_n}} = \frac{X_n - n}{\sqrt{n}}.$$

Using characteristic functions of Poisson distributions, it is easy to find the limiting distributions of random variables W_n.

Exercise 9.3 Let $g_n(t)$ be the characteristic function of W_n. Then, prove that for any fixed t,

$$g_n(t) \to e^{-t^2/2} \qquad \text{as } n \to \infty. \tag{9.33}$$

Remark 9.2 As is already known, (9.33) means that the standard normal distribution is the limiting distribution for the sequence of Poisson random variables $W_1, W_2, \ldots$. At the same time, the Poisson distribution itself is a limiting form for the binomial and some other distributions, as noted in the preceding chapters.

9.11 Mixtures

In Chapter 5 we considered the distribution of binomial $B(N, p)$ random variables in the case when N itself is a binomial random variable. Now we discuss the distribution of $B(N, p)$ random variable when N has the Poisson distribution. Later, we deal with a more general Poisson mixtures of random variables.

Let $X_1, X_2, \ldots$ be independent and identically distributed random variables having a common generating function $P(s) = \sum_{k=0}^{\infty} p_k s^k$, where $p_k = P\{X_m = k\}$, $m = 1, 2, \ldots$, $k = 0, 1, 2, \ldots$. Let $S_0 = 0$ and $S_n = X_1 + X_2 + \cdots + X_n$ ($n = 1, 2, \ldots$) be the cumulative sums of X's. Then, the generating function of S_n has the form

$$\begin{aligned} P_n(s) = Es^{S_n} &= Es^{X_1 + \cdots + X_n} \\ &= Es^{X_1} Es^{X_2} \cdots Es^{X_n} = P^n(s), \qquad n = 0, 1, \ldots. \end{aligned}$$

Consider now an integer-valued random variable N taking on values $0, 1, 2, \ldots$ with probabilities

$$q_n = P\{N = n\}, \qquad n = 0, 1, \ldots.$$

Suppose that N is independent of $X_1, X_2, \ldots$. Further, suppose that $Q(s) = \sum_{n=0}^{\infty} q_n s^n$ is the generating function of N. Let us introduce now a new random variable,

$$Y = S_N.$$

The distribution of Y is clearly a mixture of distributions of S_n taken with probabilities q_n.

Let us find probabilities $r_m = P\{Y = m\}$, $m = 0, 1, \ldots$. Due to the theorem of total probability, we readily have

$$
\begin{aligned}
r_m \;&=\; P\{Y = m\} = P\{S_N = m\} \\
&=\; \sum_{n=0}^{\infty} P\{S_N = m | N = n\} P\{N = n\} \\
&=\; \sum_{n=0}^{\infty} P\{S_n = m | N = n\} q_n.
\end{aligned}
\tag{9.34}
$$

Since random variables $S_n = X_1 + \cdots + X_n$ and N are independent,

$$P\{S_n = m | N = n\} = P\{S_n = m\},$$

and hence we may write (9.34) as

$$r_m = \sum_{n=0}^{\infty} P\{S_n = m\}\, q_n, \qquad m = 0, 1, \ldots. \tag{9.35}$$

Then, the generating function of Y has the form

$$
\begin{aligned}
R(s) \;&=\; E s^Y = \sum_{m=0}^{\infty} r_m s^m \\
&=\; \sum_{m=0}^{\infty} \sum_{n=0}^{\infty} P\{S_n = m\} q_n s^m \\
&=\; \sum_{n=0}^{\infty} q_n \sum_{m=0}^{\infty} P\{S_n = m\} s^m.
\end{aligned}
\tag{9.36}
$$

We note that the sum

$$\sum_{m=0}^{\infty} P\{S_n = m\} s^m$$

is simply the generating function of S_n and, as noted earlier, it is equal to $P^n(s)$. This enables us to simplify the RHS of (9.36) and write

$$R(s) = \sum_{n=0}^{\infty} q_n P^n(s) = Q(P(s)). \tag{9.37}$$

Relation (9.37) gives the generating function of Y, which provides a way to find the probabilities $r_0, r_1, \ldots$.

Suppose that we take $V_n \sim B(n, p)$, and we want to find the distribution of $Y = V_N$, where N is the Poisson $\pi(\lambda)$ random variable, which is independent of $V_1, V_2, \ldots$. We can then apply relation (9.37) to find the generating function of Y. In fact, for any $n = 1, 2, \ldots$, due to the properties of the binomial distribution, we have

$$V_n \overset{d}{=} X_1 + X_2 + \cdots + X_n,$$

where $X_1, X_2, \ldots$ are independent Bernoulli $Be(p)$ random variables. Therefore, we can consider Y as a sum of N independent Bernoulli random variables

$$Y \overset{d}{=} X_1 + X_2 + \cdots + X_N. \tag{9.38}$$

In this case,

$$P(s) = Es^{X_k} = 1 - p + ps, \qquad k = 1, 2, \ldots,$$

and therefore,

$$Q(s) = Es^N = e^{\lambda(s-1)}.$$

Hence, we readily obtain from (9.37) that

$$R(s) = Es^Y = Q(P(s)) = \exp\{\lambda(ps - p)\} = \exp\{\lambda p(s - 1)\}. \tag{9.39}$$

Clearly, the RHS of (9.39) is the generating function of the Poisson $\pi(\lambda p)$ random variable, so we have

$$P\{Y = m\} = \frac{e^{-\lambda p}(\lambda p)^m}{m!}, \qquad m = 0, 1, 2, \ldots.$$

Exercise 9.4 For any $n = 1, 2, \ldots$, let the random variable X_n take on values $0, 1, \ldots, n - 1$ with equal probabilities $1/n$, and let N be a random variable independent of X's having a Poisson $\pi(\lambda)$ distribution. Then, find the generating function of $Y = X_N$ and hence $P\{Y = 0\}$.

More generally, when N is distributed as Poisson with parameter λ so that $Q(s) = e^{\lambda(s-1)}$ [see (9.2)], (9.37) becomes

$$R(s) = e^{\lambda\{P(s)-1\}}.$$

This then is the generating function of the distribution of the sum of a Poisson number of i.i.d. random variables with generating function $P(s)$. The corresponding distributions are called *Poisson-stopped-sum distributions*, a name introduced by Godambe and Patil (1975) and adopted since then by a number of authors including Douglas (1980) and Johnson, Kotz, and Kemp (1992). Some other names such as *generalized Poisson* [Feller (1943)], *stuttering Poisson* [Kemp (1967)] and *compound Poisson* [Feller (1968)] have also been used for these distributions.

Exercise 9.5 Show that the negative binomial distribution in (7.2) is a Poisson-stopped-sum distribution with $P(s)$ being the generating function of a *logarithmic distribution* with mass function

$$
p_k = \begin{cases} 1 + \dfrac{\alpha}{\lambda} \log(1 - p) & \text{for } k = 0 \\[2mm] \dfrac{\alpha}{\lambda} \dfrac{p^k}{k} & \text{for } k = 1, 2, \ldots . \end{cases}
$$

9.12   Rao–Rubin Characterization

In this section we present (without proof) the following celebrated *Rao–Rubin characterization* of the Poisson distribution [Rao and Rubin (1964)]. If X is a discrete random variable taking on only nonnegative integral values and that the conditional distribution of Y, given $X = x$, is binomial $B(x, p)$ (where p does not depend on x), then the distribution of X is Poisson if and only if

$$
P\left(Y = y | Y = X\right) = P\left(Y = y | Y \neq X\right). \tag{9.40}
$$

An interesting physical interpretation of this result was given by Rao (1965) wherein X represents a naturally occurring quantity with some of its components not being counted (or destroyed) when it is observed, and Y represents the value remaining (that is, the components of X which are actually counted) after this destructive process. In other words, suppose that X is the original observation having a Poisson $\pi(\lambda)$ distribution, and the probability that the original observation n gets reduced to x due to the destructive process is

$$
\binom{n}{y} p^y (1 - p)^{n - y}, \qquad y = 0, 1, \ldots, n, \qquad 0 < p < 1. \tag{9.41}
$$

Now, if Y represents the resulting random variable, then

$$
\begin{aligned} P(Y = y) &= P\left(Y = y | \text{destroyed}\right) = P\left(Y = y | \text{not destroyed}\right) \\ &= \frac{e^{-p\lambda}(p\lambda)^y}{y!}; \end{aligned} \tag{9.42}
$$

furthermore, the condition in (9.42) also characterizes the Poisson distribution.

Interestingly, Srivastava and Srivastava (1970) established that if the original observations follow a Poisson distribution and if the condition in (9.42) is satisfied, then the destructive process has to be binomial, as in (9.41).

The Rao–Rubin characterization result above generated a lot of interest, which resulted in a number of different variations, extensions, and generalizations.

9.13 Generalized Poisson Distribution

Let X be a random variable defined over nonnegative integers with its probability function as

$$
\begin{aligned}
p_m(\theta, \lambda) &= P\{X = m\} \\
&= \begin{cases} \dfrac{1}{m!}\, \theta(\theta + m\lambda)^{m-1} e^{-\theta - m\lambda}, & m = 0, 1, \ldots, \\ 0 & \text{for } m > \ell \text{ when } \lambda < 0, \end{cases}
\end{aligned} \tag{9.43}
$$

where $\theta > 0$, $\max(-1, -\theta/\ell) < \lambda \le 1$, and $\ell\ (\ge 4)$ is the largest positive integer for which $\theta + \ell\lambda > 0$ when λ is negative. Then, X is said to have the *generalized Poisson* (GPD) *distribution*. A book-length account of generalized Poisson distributions, discussing in great detail their various properties and applications, is available and is due to Consul (1989). This distribution, also known as the *Lagrangian–Poisson distribution*, is a Poisson-stopped-sum distribution.

The special case when $\lambda = \alpha\theta$ in (9.43) is called the *restricted generalized Poisson distribution*, and the probability function in this case becomes

$$
p_m(\theta, \lambda) = P\{X = m\} = \frac{1}{m!}\, \theta^m (1 + m\alpha)^{m-1} e^{-\theta - m\alpha\theta}, \qquad m = 0, 1, \ldots, \tag{9.44}
$$

where $\max(-1/\theta, -1/\ell) < \alpha < 1/\theta$.

Exercise 9.6 For the restricted generalized Poisson distribution in (9.44), show that the mean and variance are given by

$$
\frac{\theta}{1 - \alpha\theta} \quad \text{and} \quad \frac{\theta}{(1 - \alpha\theta)^3}, \tag{9.45}
$$

respectively.

CHAPTER 10

MISCELLANEA

10.1 Introduction

In the last eight chapters we have seen the most popular and commonly encountered discrete distributions. There are a few natural generalizations of some of these distributions which are not only of interest due to their mathematical niceties but also due to their interesting probabilistic basis. These are described briefly in this chapter, and their basic properties are also presented.

10.2 Pólya Distribution

Pólya (1930) suggested this distribution in the context of the following combinatorial problem; see also Eggenberger and Pólya (1923). There is an urn with b black and r red balls. A ball is drawn at random, after which $c + 1$ (where $c \geq -1$) new balls of the same color are added to the urn. We repeat this process successively n times. Let X_n be the total number of black balls observed in these n draws. It can easily be shown that X_n takes on values $0, 1, \ldots, n$ with the following probabilities:

$$p_k = \binom{n}{k} \frac{b(b+c) \cdots \{b + (k-1)c\} r(r+c) \cdots \{r + (n-k-1)c\}}{(b+r)(b+r+c)(b+r+2c) \cdots \{b+r+(n-1)c\}}, \quad (10.1)$$

where $k = 0, 1, \ldots, n$ and $n = 1, 2, \ldots$. Let us now denote

$$p = \frac{b}{b+r}, \quad q = 1 - p = \frac{r}{b+r}, \quad \alpha = \frac{c}{b+r}. \quad (10.2)$$

Then, the probability expression in (10.1) can be rewritten in the form

$$p_k = \binom{n}{k} \frac{p(p+\alpha) \cdots \{p + (k-1)\alpha\} q(q+\alpha) \cdots \{q + (n-k-1)\alpha\}}{(1+\alpha)(1+2\alpha) \cdots \{1 + (n-1)\alpha\}}, \quad (10.3)$$

wherein we can forget that p, q, and α are quotients of integers and suppose only that $0 < p, q < 1$, and $\alpha > -1/(n-1)$. Note that if $\alpha > 0$, then (10.3), for $1 < k < n$, can be expressed in terms of complete beta functions as

$$p_k = \frac{nB\left(1/\alpha, n\right)}{k(n-k)B\left(p/\alpha, k\right)B\left(q/\alpha, n-k\right)}, \tag{10.4}$$

where the complete beta function is given by

$$B(a, b) = \int_0^1 t^{a-1}(1-t)^{b-1}\, dt, \qquad a > 0, \quad b > 0.$$

Distribution in (10.3) is called the *Pólya distribution* with parameters n, p, and α, where $n = 1, 2, \ldots$, $0 < p < 1$ and $\alpha > -1/(n-1)$.

If $\alpha = 0$, then the probabilities in (10.3) simply become the binomial $B(n, p)$ probabilities. As a matter of fact, in the Pólya urn scheme, we see that this case corresponds to the number of black balls in a sample of size n with replacement from the urn, in which the ratio of black balls equals p.

Next, we know from Chapter 8 that a hypergeometric random variable can be interpreted as the number of black balls in a sample of size n without replacement from the urn. Clearly, this corresponds to the Pólya urn scheme with $c = -1$, that is, $\alpha = -1/(b+r)$.

Thus, Pólya distributions include binomial as well as hypergeometric distributions as special cases, and hence may be considered as a "bridge" between these two types of distributions. Note also that if a random variable X has distribution (10.4), then

$$EX = np \tag{10.5}$$

and

$$\text{Var } X = np(1-p)\,\frac{1+\alpha n}{1+\alpha}. \tag{10.6}$$

10.3 Pascal Distribution

From Chapter 7 we know that the negative binomial $NB(m, p)$ distribution with integer parameter m is same as the distribution of the sum of m independent geometrically distributed $G(p)$ random variables. If $X \sim NB(m, p)$, then

$$P\{X = n\} = \binom{n+m-1}{n-1}(1-p)^m p^n, \qquad n = 1, 2, \ldots. \tag{10.7}$$

The appearance of this class of distributions in this manner was, therefore, quite natural. Later, the family of distributions (10.7) was enlarged to the family of negative binomial distributions $NB(\alpha, p)$ for any positive parameter α. Negative binomial distributions with integer parameter α are sometimes (especially if we want to emphasize that α is an integer) called the *Pascal distributions*.

10.4 Negative Hypergeometric Distribution

The Pascal distribution (i.e., the negative binomial distribution with integer parameter α) has the following "urn and balls" interpretation. In an urn that contains black and red balls, suppose that the proportion of black balls is p. Balls are drawn from the urn with replacement, which means that we have the Pólya urn scheme with $c = 0$. The sampling process is considered to be completed when we get the mth red ball (i.e., the mth "failure"). Then, the number of black balls (i.e., the number of "successes") in our random sample has the distribution given in (10.7).

Let us now consider an urn having r red and b black balls. Balls are drawn without replacement (i.e., the Pólya urn scheme with $c = -1$). The sampling process is once again considered to be completed when we get the mth red ball. Let X be the number of black balls in our sample. Then, it is not difficult to see that X takes on values $0, 1, \ldots, b$ with probabilities

$$p_k = P\{X = k\} = \frac{\binom{k + m - 1}{m - 1}\binom{b + r - k - m}{r - m}}{\binom{b + r}{r}}, \qquad 0 \le k \le b. \quad (10.8)$$

Such a random variable X is said to have the *negative hypergeometric distribution* with parameters $b = 1, 2, \ldots$, $r = 1, 2, \ldots$, and $m = 1, 2, \ldots, r$.

Exercise 10.1 Show that the mean and variance of the random variable above are given by

$$EX = \frac{mb}{r + 1} \qquad (10.9)$$

and

$$\text{Var } X = \frac{m(b + r + 1)b(r - m + 1)}{(r + 1)^2(r + 2)}, \qquad (10.10)$$

respectively.

Part II

CONTINUOUS DISTRIBUTIONS

CHAPTER 11

UNIFORM DISTRIBUTION

11.1 Introduction

The uniform distribution is the simplest of all continuous distributions, yet is one of the most important distributions in continuous distribution theory. As shown in Chapter 2, the uniform distribution arises naturally as a limiting form of discrete uniform distributions.

11.2 Notations

We say that the random variable X has the *standard uniform distribution* if its pdf is of the form

$$p_X(x) = \begin{cases} 1 & \text{if } 0 \leq x \leq 1 \\ 0 & \text{otherwise.} \end{cases} \tag{11.1}$$

The corresponding cdf is given by

$$F_X(x) = \begin{cases} 0 & \text{if } x < 0 \\ x & \text{if } 0 \leq x \leq 1 \\ 1 & \text{if } x > 1. \end{cases} \tag{11.2}$$

We use the notation $X \sim U(0,1)$ for the standard uniform distribution, concentrated on the interval $[0,1]$. As we well know, the linear transformation $Y = a + hX$, $h > 0$, preserves the type of distributions and gives us a new random variable Y, which takes on values in the interval $[a, a+h]$. It is easy to see that

$$\begin{aligned} F_Y(x) &= P\{Y \leq x\} = P\{a + hX \leq x\} \\ &= P\left\{ X \leq \frac{x-a}{h} \right\} \\ &= F_X\left(\frac{x-a}{h} \right) \\ &= \frac{x-a}{h} \qquad \text{if } a \leq x \leq a+h, \end{aligned} \tag{11.3}$$

and the corresponding pdf is of the form

$$p_Y(x) = \begin{cases} \dfrac{1}{h} & \text{if } a \le x \le a+h \\ 0 & \text{otherwise.} \end{cases} \tag{11.4}$$

For $h = b - a$ $(b > a)$, we obtain the distribution with pdf

$$p(x) = \begin{cases} \dfrac{1}{b-a} & \text{if } a \le x \le b \\ 0 & \text{otherwise,} \end{cases} \tag{11.5}$$

and cdf

$$F(x) = \begin{cases} 0 & \text{if } x < a \\ \dfrac{x-a}{b-a} & \text{if } a \le x \le b \\ 1 & \text{if } x > b. \end{cases} \tag{11.6}$$

We denote this distribution by $U(a,b)$, and call it the *uniform $U(a,b)$ distribution*. Some authors call it the *rectangular distribution*.

11.3 Moments

Since the random variable $X \sim U(a,b)$ has a finite support, all its moments exist.

Moments about zero:
   If $X \sim U(a,b)$, then

$$\begin{aligned} \alpha_n = EX^n &= \int_a^b x^n \frac{1}{b-a}\, dx \\ &= \frac{b^{n+1} - a^{n+1}}{(n+1)(b-a)} \end{aligned} \tag{11.7}$$

and, in particular, we have the first four moments as

$$\alpha_1 = EX = \frac{a+b}{2}, \tag{11.8}$$

$$\alpha_2 = EX^2 = \frac{a^2 + ab + b^2}{3}, \tag{11.9}$$

$$\alpha_3 = EX^3 = \frac{a^3 + a^2 b + ab^2 + b^3}{4}, \tag{11.10}$$

$$\alpha_4 = EX^4 = \frac{a^4 + a^3 b + a^2 b^2 + ab^3 + b^4}{5}, \tag{11.11}$$

respectively.
   From (11.7), we also note that

$$\alpha_n = \frac{1}{n+1}, \quad n = 1, 2, \ldots, \tag{11.12}$$

for the standard uniform $U(0,1)$ distribution.

Central moments:
    For any $n = 1, 2, \ldots$, we obtain from (11.7) that

$$
\begin{aligned}
\beta_n &= E\left(X - EX\right)^n = \frac{1}{b-a} \int_a^b \left(x - \frac{a+b}{2}\right)^n dx \\
&= \frac{1}{b-a} \int_{(a-b)/2}^{(b-a)/2} v^n \, dv \\
&= \frac{(b-a)^n \left\{1 - (-1)^{n+1}\right\}}{(n+1)\, 2^{n+1}}.
\end{aligned}
\tag{11.13}
$$

Thus, we simply obtain for any $n = 1, 2, \ldots$,

$$
\begin{aligned}
\beta_{2n-1} &= 0, \\
\beta_{2n} &= \frac{(b-a)^{2n}}{(2n+1)\, 4^n}.
\end{aligned}
\tag{11.14}
$$

From (11.14), we readily have the variance of X to be

$$
\operatorname{Var} X = \beta_2 = \frac{(b-a)^2}{12},
\tag{11.15}
$$

and, in particular,

$$
\operatorname{Var} X = \frac{1}{12}
$$

for the standard uniform $U(0,1)$ distribution.

Exercise 11.1 Let X and Y be independent random variables with uniform $U(0,1)$ distribution. Then, find the cumulative distribution function, probability density function, and expectation of the random variable $Z = XY$.

Shape characteristics:
    From (11.14), we note that the uniform $U(a,b)$ distribution has its Pearson coefficient of skewness γ_1 to be zero. In fact, in this case, the distribution is also symmetric about the mean $(a+b)/2$. Furthermore, we obtain the coefficient of kurtosis as

$$
\gamma_2 = \frac{\beta_4}{\beta_2^2} = \frac{9}{5} = 1.8.
\tag{11.16}
$$

Hence, the uniform $U(a,b)$ distribution is symmetric platykurtic for any choice of a and b.

11.4   Entropy

Recall from Chapter 1 that if a random variable X has pdf $p(x)$, then its entropy is defined as

$$H(X) = -\int_{\mathcal{D}} p(x) \log p(x) \, dx, \tag{11.17}$$

where $\mathcal{D} = \{x : p(x) > 0\}$. Now, if $X \sim U(a, b)$, then

$$H(X) = -\int_a^b \frac{1}{b-a} \log\left(\frac{1}{b-a}\right) dx = \log(b-a). \tag{11.18}$$

Remark 11.1 It is of interest to note that among all the absolutely continuous distributions having support $[a, b]$, the maximal value of the entropy is attained by the uniform $U(a, b)$ distribution.

11.5   Characteristic Function

If $X \sim U(a, b)$, then its characteristic function has the form

$$
\begin{aligned}
f_X(t) &= Ee^{itX} = \frac{1}{b-a} \int_a^b e^{itx} \, dx \\
&= \frac{e^{itb} - e^{ita}}{(b-a)it} .
\end{aligned}
\tag{11.19}
$$

In particular, we have

$$f_X(t) = \frac{e^{it} - 1}{it} \tag{11.20}$$

for the standard uniform $U(0, 1)$ distribution, and

$$f_X(t) = \frac{e^{it} - e^{-it}}{2} = \frac{\sin t}{t} \tag{11.21}$$

for the symmetric uniform $U(-1, 1)$ distribution.

11.6   Convolutions

Let X_1 and X_2 be independent random variables with pdf's $p_1(x)$ and $p_2(x)$, respectively. To find the pdf $p_Y(x)$ of the sum $Y = X_1 + X_2$, we may use the following equalities:

$$
\begin{aligned}
p_Y(x) &= \int_{-\infty}^{\infty} p_{X_1}(x - y) p_{X_2}(y) \, dy \\
&= \int_{-\infty}^{\infty} p_{X_2}(x - y) p_{X_1}(y) \, dy.
\end{aligned}
\tag{11.22}
$$

Exercise 11.2 Suppose that $X_1 \sim U(0,1)$ and $X_2 \sim U(0,1)$, and that they are independent. Show then that the pdf of $Y = X_1 + X_2$ is given by

$$p_Y(x) = \begin{cases} x & \text{if } 0 \le x \le 1 \\ 2 - x & \text{if } 1 \le x \le 2 \\ 0 & \text{otherwise.} \end{cases} \tag{11.23}$$

Remark 11.2 The distribution with pdf (11.23) is called the *triangular distribution*, and the shape of $p_Y(x)$ gives the obvious reason for this name.

The pdf of the sum $Y_n = X_1 + X_2 + \cdots + X_n$ of independent $U(0,1)$ distributed random variables $X_1, X_2, \ldots, X_n$ has the following more complicated form:

$$p_{Y_n}(x) = \frac{1}{(n-1)!} \sum_{j=0}^{k} (-1)^j \binom{n}{j} (x-j)^{n-1} \tag{11.24}$$

if $k \le x \le k+1$ for $k = 0, 1, \ldots, n-1$, and 0 otherwise. Indeed, for the case when $n = 2$, (11.24) reduces to the simple expression in (11.23).

11.7   Decompositions

Let $X \sim U(0,1)$, $V \sim U(0, \frac{1}{2})$, and $Y \sim Be(\frac{1}{2})$, where V and Y are independent random variables. It then turns out that

$$X \stackrel{d}{=} V + \frac{Y}{2} . \tag{11.25}$$

In fact, the characteristic functions of X, V, and $Y/2$ have the following form:

$$\begin{aligned} f_X(t) &= \frac{1}{it}(e^{it} - 1), \\ f_V(t) &= \frac{2}{it}(e^{\frac{it}{2}} - 1), \\ f_{\frac{Y}{2}}(t) &= \frac{1}{2}(e^{\frac{it}{2}} + 1). \end{aligned}$$

Since

$$f_X(t) = f_V(t) f_{\frac{Y}{2}}(t),$$

relation (11.25) follows immediately.

Developing this idea further, we can use the equality

$$\frac{e^{it} - 1}{it} = \frac{2^n \left\{ \exp\left(\frac{it}{2^n}\right) - 1 \right\}}{it} \times \frac{\exp\left(\frac{it}{2^n} + 1\right)}{2} \times \cdots \times \frac{\exp\left(\frac{it}{2} + 1\right)}{2} \tag{11.26}$$

to show that for any $n = 1, 2, \ldots$,

$$X \stackrel{d}{=} V_n + \frac{Y_1}{2} + \cdots + \frac{Y_n}{2^n}, \tag{11.27}$$

where $X \sim U(0, 1)$, $V_n \sim U(0, 2^{-n})$, and $Y_1, Y_2, \ldots$ all have Bernoulli $Be(\frac{1}{2})$ distribution, and all the random variables on the RHS of (11.27) are independent.

We thus see that $X \sim U(0, 1)$ can be expressed as a sum of any number of independent random variables. Yet its distribution is not infinitely divisible, since X has a finite support.

11.8 Probability Integral Transform

Consider a random variable Y having a continuous cdf F. Let G be the inverse of F, defined as

$$G(u) = \inf\{x : F(x) \geq u\}, \quad 0 < u < 1. \tag{11.28}$$

Note that $F(G(x)) = x$ for $0 < x < 1$.

Let us now introduce a new random variable,

$$X = F(Y).$$

It is then clear that $0 \leq X \leq 1$. Also, it is easy to find the cdf of X as

$$
\begin{aligned}
F_X(x) &= P\{X \leq x\} = P\{F(Y) \leq x\} \\
&= P\{Y \leq G(x)\} = F(G(x)) = x \quad \text{for} \ \ 0 < x < 1.
\end{aligned}
$$

This simply means that $X \sim U(0, 1)$. Hence, for any random variable Y with a continuous cdf F, the transformation $X = F(Y)$ results in a standard uniform $U(0, 1)$ random variable.

On the other hand, let F be any distribution function and let G be its inverse as defined in (11.28). Then, it turns out that the transformation $Y = G(X)$, where $X \sim U(0, 1)$, will yield a random variable Y with the prefixed cdf F.

Remark 11.3 This property is very useful in simulating samples from any continuous distribution F, as the problem can be reduced to simulation from standard uniform $U(0, 1)$ distribution. Thus, with the aid of an efficient uniform random number generator and a method to determine G, we can easily construct an algorithm for simulating samples from F.

11.9 Distributions of Minima and Maxima

Consider a sequence of independent random variables $X_1, X_2, \ldots$ having the standard uniform $U(0, 1)$ distribution. Let

$$m_n = \min\{X_1, X_2, \ldots, X_n\} \tag{11.29}$$

and

$$M_n = \max\{X_1, X_2, \ldots, X_n\}, \qquad n = 1, 2, \ldots. \tag{11.30}$$

It is then easy to see that for $0 \le x \le 1$,

$$\begin{aligned}
G_n(x) &= P\{m_n \le x\} = 1 - P\{m_n > x\} \\
&= 1 - P\{X_1 > x, \ldots, X_n > x\} \\
&= 1 - (1-x)^n \tag{11.31}
\end{aligned}$$

and

$$H_n(x) = P\{M_n \le x\} = P\{X_1 \le x, \ldots, X_n \le x\} = x^n. \tag{11.32}$$

Note that

$$EM_n = \int_0^1 x \, d(x^n) = \frac{n}{n+1}, \tag{11.33}$$

$$Em_n = -\int_0^1 x \, d\{(1-x)^n\} = \frac{1}{n+1}, \tag{11.34}$$

and

$$1 - EM_n = Em_n = \frac{1}{n+1} \to 0 \quad \text{as } n \to \infty. \tag{11.35}$$

This shows that for large n, $1 - M_n$ and m_n are close to zero. To examine the behavior of m_n and $1 - M_n$ for large n, we will now determine the asymptotic distributions of these random variables, after suitable normalization. It turns out that

$$\begin{aligned}
G_n\left(\frac{x}{n}\right) &= P\left\{m_n \le \frac{x}{n}\right\} = P\{n \, m_n \le x\} \\
&= 1 - \left(1 - \frac{x}{n}\right)^n \to 1 - e^{-x} \quad \text{for } x > 0, \tag{11.36}
\end{aligned}$$

as $n \to \infty$. Thus, we have established that the asymptotic distribution function of $n \min(X_1, X_2, \ldots, X_n)$ is given by

$$G(x) = \begin{cases} 0 & \text{if } x < 0 \\ 1 - e^{-x} & \text{if } x \ge 0. \end{cases} \tag{11.37}$$

This is the *exponential distribution*, which is discussed in detail in Chapter 18.

Exercise 11.3 Show that

$$P\{n(M_n - 1) < x\} \to e^x \qquad \text{for } x < 0 \tag{11.38}$$

as $n \to \infty$.

Remark 11.4 Consider the distribution functions

$$R_n(x) = 1 - \left(1 - \frac{x}{n}\right)^n, \qquad 0 \leq x \leq n, \ n = 1, 2, \ldots$$

which we encountered in (11.36). Following Proctor (1987), we can say that they form the family of *generalized uniform distributions* as the sequence $R_1(x), R_2(x), \ldots$ forms a "bridge" between the uniform distribution $[R_1(x) = x, \ 0 \leq x \leq 1]$ and the exponential distribution $[R_\infty(x) = \lim_{n \to \infty} R_n(x) = 1 - e^{-x}, \ x \geq 0]$.

11.10   Order Statistics

Let us consider n independent standard uniform $U(0, 1)$ random variables $X_1, X_2, \ldots, X_n$. In Section 11.9 we investigated the maximal and minimal values of $X_1, \ldots, X_n$. We can similarly consider other values among the ordering of $X_1, \ldots, X_n$. A result of such an ordering gives the *order statistics* (in this case, uniform order statistics)

$$X_{1,n} \leq X_{2,n} \leq \cdots \leq X_{n,n}. \tag{11.39}$$

In particular,

$$X_{1,n} = \min\{X_1, \ldots, X_n\}$$

and

$$X_{n,n} = \max\{X_1, \ldots, X_n\},$$

which were denoted earlier by m_n and M_n, respectively.

Let

$$F_{k,n}(x) = P\{X_{k,n} \leq x\} \qquad \text{for } 0 \leq x \leq 1, \ 1 \leq k \leq n, \ n = 1, 2, \ldots,$$

denote the cdf of the order statistic $X_{k,n}$. Simple probability arguments yield an expression for $F_{k,n}(x)$. In fact, the event $A_{k,n}^{(x)} = \{X_{k,n} \leq x\}$ coincides with the event $\bigcup_{r=k}^{n} B_m(x)$, where

$$B_r(x) = \{\text{exactly } r \text{ of the variables } X_1, \ldots, X_n \text{ are at most } x\}.$$

It is then easy to see that

$$P\{B_r(x)\} = \binom{n}{r} x^r (1 - x)^{n-r}$$

and, consequently,

$$P\{A_{k,n}^{(x)}\} = \sum_{r=k}^{n} \binom{n}{r} x^r (1-x)^{n-r}. \tag{11.40}$$

Using the equality in (5.36), we readily obtain for $0 \le x \le 1$,

$$\begin{aligned}
F_{k,n}(x) &= P\{A_{k,n}^{(x)}\} \\
&= \sum_{r=k}^{n} \binom{n}{r} x^r (1-x)^{n-r} \\
&= \frac{n!}{(k-1)!\,(n-k)!} \int_0^x t^{k-1}(1-t)^{n-k}\, dt. \tag{11.41}
\end{aligned}$$

Note that the cdf given in (11.41) belongs to the family of *beta distributions*, which is discussed in detail in Chapter 16.

Upon differentiating (11.41) with respect to x, we immediately obtain the pdf of $X_{k,n}$ as

$$p_{k,n}(x) = \frac{n!}{(k-1)!\,(n-k)!} x^{k-1}(1-x)^{n-k}, \qquad 0 \le x \le 1. \tag{11.42}$$

From (11.42), we readily find the mean and variance of $X_{k,n}$ as

$$EX_{k,n} = \frac{k}{n+1} \tag{11.43}$$

and

$$\operatorname{Var} X_{k,n} = \frac{k(n-k+1)}{(n+1)^2(n+2)}. \tag{11.44}$$

Exercise 11.4 Derive the formulas presented in (11.43) and (11.44).

Using similar probability arguments, we can derive the joint pdf of $X_{k,n}$ and $X_{\ell,n}$ (for $1 \le k < \ell \le n$). For this purpose, let

$$F_{k,\ell,n}(x,y) = P\{X_{k,n} \le x, X_{\ell,n} \le y\} \quad \text{for } 0 \le x < y \le 1,\ 1 \le k < \ell \le n$$

denote the joint cdf of the order statistics $X_{k,n}$ and $X_{\ell,n}$. We then observe that for $0 \le x < y \le 1$,

$$\begin{aligned}
F_{k,\ell,n}(x,y) &= P\{\text{at least } k \text{ of } X_1,\ldots,X_n \text{ are at most } x \text{ and} \\
&\qquad \text{at least } \ell \text{ of } X_1,\ldots,X_n \text{ are at most } y\} \\
&= \sum_{s=\ell}^{n} \sum_{r=k}^{s} \frac{n!}{r!\,(s-r-1)!\,(n-s)!}\, x^r (y-x)^{s-r}(1-y)^{n-s}.
\end{aligned}$$

$$\tag{11.45}$$

Upon using the identity that the RHS of (11.45) equals

$$\frac{n!}{(k-1)!\,(\ell-k-1)!\,(n-\ell)!}\int_0^x\int_{t_1}^y t_1^{k-1}(t_2-t_1)^{\ell-k-1}$$
$$\times(1-t_2)^{n-\ell}\,dt_2\,dt_1, \qquad (11.46)$$

we obtain the joint density of $X_{k,n}$ and $X_{\ell,n}$ for $0 \le x < y \le 1$ and $1 \le k < \ell \le n$ as

$$F_{k,l,n}(x,y) = \frac{n!}{(k-1)!\,(\ell-k-1)!\,(n-\ell)!}\int_0^x\int_{t_1}^y t_1^{k-1}$$
$$\times(t_2-t_1)^{\ell-k-1}(1-t_2)^{n-\ell}\,dt_2\,dt_1. \qquad (11.47)$$

Upon differentiating (11.47) with respect to y and x, we obtain the joint pdf of $X_{k,n}$ and $X_{\ell,n}$ as

$$p_{k,\ell,n}(x,y) = \frac{n!}{(k-1)!\,(\ell-k-1)!\,(n-\ell)!}x^{k-1}(y-x)^{\ell-k-1}(1-y)^{n-\ell},$$
$$0 \le x < y \le 1. \qquad (11.48)$$

Exercise 11.5 Prove the combinatorial identity in (11.46).

From (11.48) we can show that

$$E\left(X_{k,n}X_{\ell,n}\right) = \frac{k(\ell+1)}{(n+1)(n+2)},$$

so that the covariance between $X_{k,n}$ and $X_{\ell,n}$ is given by

$$\begin{aligned}
\mathrm{Cov}\left(X_{k,n},X_{\ell,n}\right) &= E\left(X_{k,n}X_{\ell,n}\right) - EX_{k,n}\,EX_{\ell,n} \\
&= \frac{k(\ell+1)}{(n+1)(n+2)} - \frac{k\ell}{(n+1)^2} \\
&= \frac{k(n-\ell+1)}{(n+1)^2(n+2)}. \qquad (11.49)
\end{aligned}$$

Furthermore, combining the formulas in (11.44) and (11.49), we obtain the correlation coefficient between $X_{k,n}$ and $X_{\ell,n}$ for $1 \le k < \ell \le n$ as

$$\rho(X_{k,n},X_{\ell,n}) = \sqrt{\frac{k(n-\ell+1)}{\ell(n-k+1)}}. \qquad (11.50)$$

Exercise 11.6 Derive the formulas presented in (11.49) and (11.50).

Remark 11.5 Let $Y_1, \ldots, Y_n$ be a random sample from a continuous pdf $p(y)$ and cdf $F(y)$, and $Y_{1,n} \leq Y_{2,n} \leq \cdots \leq Y_{n,n}$ denote the corresponding order statistics. Since $F(Y_i) \stackrel{d}{=} X_i$, where $X_1, \ldots, X_n$ is a random sample from the standard uniform $U(0,1)$ distribution (see Section 11.8) and that the transformation is order-preserving, we readily obtain from (11.42) the pdf of $Y_{k,n}$ as

$$p_{Y_{k,n}}(x) = \frac{n!}{(k-1)!\,(n-k)!} \{F(x)\}^{k-1} \{1 - F(x)\}^{n-k} p(x),$$
$$-\infty < x < \infty, \qquad (11.51)$$

and from (11.48) the joint pdf of $Y_{k,n}$ and $Y_{\ell,n}$ $(1 \leq k < \ell \leq n)$ as

$$p_{Y_{k,n},Y_{\ell,n}}(x,y) = \frac{n!}{(k-1)!\,(\ell-k-1)!\,(n-\ell)!} \{F(x)\}^{k-1}$$
$$\times \{F(y) - F(x)\}^{\ell-k-1} \{1 - F(y)\}^{n-\ell} p(x)p(y),$$
$$-\infty < x < y < \infty. \qquad (11.52)$$

11.11   Relationships with Other Distributions

We have already seen the relationship of uniform distribution to discrete uniform, triangular, exponential, and beta distributions. Now, we will show that Cauchy distributions are closely related to the uniform distribution.

Let $O = (0,0)$, $A = (1,0)$, $E = (0,-1)$ and $D = (0,1)$. Let the point B be chosen uniformly on the semicircle DAE meaning that θ, the angle between OE and OB, has the uniform $U(-\pi/2, \pi/2)$ distribution. Let $C = (1, Y)$ be the point of contact of the extension of the line OB on the line $Y = 1$. The second coordinate of C is a random variable Y which is negative if $-\pi/2 < \theta < 0$, and is positive if $0 < \theta < \pi/2$.

Exercise 11.7 Show that the cdf and pdf of Y are given by

$$P\{Y < y\} = \frac{1}{2} + \frac{1}{\pi}\tan^{-1}y, \quad p_Y(y) = \frac{1}{\pi(1+y^2)}, \quad -\infty < y < \infty. \ (11.53)$$

The distribution of Y in (11.53) is called the *Cauchy distribution*, which is discussed in detail in Chapter 12.

Let X be distributed as standard uniform $U(0,1)$. Then, Tukey (1962) studied the transformation

$$Y = \frac{aX^\lambda - (1-X)^\lambda}{\lambda}, \quad a > 0, \ \lambda \neq 0, \ -\frac{1}{\lambda} \leq Y \leq \frac{a}{\lambda} \text{ for } \lambda > 0,$$

(11.54)

with the transitional transformation (case $\lambda = 0$) being

$$Y = \log\left(\frac{X^a}{1-X}\right), \quad a > 0 .$$

(11.55)

The distribution of Y is called as *Tukey's lambda distribution*. The case $a = 1$ is referred to as *Tukey's symmetric lambda distribution*, in which case

$$Y = \frac{X^\lambda - (1-X)^\lambda}{\lambda}, \quad \lambda \neq 0,$$

(11.56)

and

$$Y = \log\left(\frac{X}{1-X}\right) \quad \text{as } \lambda \to 0.$$

(11.57)

Exercise 11.8 If Y follows Tukey's lambda distribution as in (11.54), show that its nth moment about zero is given by

$$EY^n = \frac{1}{\lambda^n} \sum_{j=0}^{n} (-1)^j \binom{n}{j} a^{n-j} B\left(\lambda(n-j) + 1, \lambda j + 1\right),$$

where $B(\cdot\, , \ \cdot)$ denotes the complete beta function. Show, in particular, that the mean and variance of Y are given by

$$EY = \frac{a-1}{\lambda(\lambda+1)}$$

and

$$\text{Var } Y = \frac{a^2 + 1}{(\lambda+1)(2\lambda+1)} - \frac{2a}{\lambda^2}\left\{ B(\lambda+1, \lambda+1) - \frac{1}{(\lambda+1)^2} \right\},$$

respectively.

CHAPTER 12

CAUCHY DISTRIBUTION

12.1   Notations

In Chapter 11, in Exercise 11.7, we found a new distribution with pdf

$$p(x) = \frac{1}{\pi(1+x^2)}, \qquad -\infty < x < \infty, \tag{12.1}$$

and cdf

$$F(x) = \frac{1}{2} + \frac{1}{\pi}\tan^{-1} x, \qquad -\infty < x < \infty. \tag{12.2}$$

A random variable X with pdf (12.1) and cdf (12.2) is said to have the *standard Cauchy distribution*. A linear transformation

$$Y = a + hX$$

gives a random variable with pdf

$$p_Y(x) = \frac{1}{\pi h \left\{ 1 + \dfrac{(x-a)^2}{h^2} \right\}} = \frac{h}{\pi \left\{ h^2 + (x-a)^2 \right\}}, \quad -\infty < x < \infty, \tag{12.3}$$

and cdf

$$F_Y(x) = \frac{1}{2} + \frac{1}{\pi}\tan^{-1}\left(\frac{x-a}{h}\right), \quad -\infty < x < \infty, \ -\infty < a < \infty, \ h > 0. \tag{12.4}$$

The random variable Y belongs to the Cauchy type of distributions, and we will denote it by

$$Y \sim C(a,h).$$

Indeed, the random variable $X \sim C(0,1)$ then has the standard Cauchy distribution defined by (12.1) and (12.2).

Note that $X \sim C(0,1)$ is expressed in terms of the uniform $U(0,1)$ random variable U as

$$X \sim \tan\left\{ \pi\left(U - \frac{1}{2}\right)\right\}. \tag{12.5}$$

12.2 Moments

Let $X \sim C(0,1)$. Then, $p_X(x) \sim 1/(\pi x^2)$ as $x \to \infty$ or $x \to -\infty$. Hence,

$$E|X|^\alpha = \int_{-\infty}^\infty |x|^\alpha p_\xi(x) \, dx = \infty$$

if $\alpha \geq 1$, and $E|X|^\alpha < \infty$ if $\alpha < 1$.

This means that the moments of order $\alpha < 1$ do exist, while the moments of order $\alpha \geq 1$ do not exist for Cauchy distributions.

12.3 Characteristic Function

Let us consider

$$f_X(t) = \int_{-\infty}^\infty \frac{e^{itx}}{\pi(1+x^2)} \, dx,$$

which is the characteristic function of the random variable $X \sim C(0,1)$. With the help of complex contour integration, it can be shown that

$$f_X(t) = e^{-|t|}. \tag{12.6}$$

Then the characteristic function of Y, having the Cauchy $C(a,h)$ distribution, becomes

$$f_Y(t) = e^{ita-h|t|}. \tag{12.7}$$

12.4 Convolutions

Let $U \sim C(a_1, h_1)$ and $V \sim C(a_2, h_2)$ be independent random variables, and let $Y = U + V$. Then the characteristic function of Y is given by

$$\begin{aligned} f_Y(t) &= f_U(t)f_V(t) = e^{ita_1-h_1|t|} \, e^{ita_2-h_2|t|} \\ &= e^{it(a_1+a_2)-(h_1+h_2)|t|}, \end{aligned} \tag{12.8}$$

which simply implies that

$$Y \sim C(a_1 + a_2, h_1 + h_2). \tag{12.9}$$

Generalizing this result, if we assume $Y_k \sim C(a_k, h_k)$ $(k = 1, 2, \ldots, n)$ to be independent random variables, and $Y = Y_1 + \cdots + Y_n$, then $Y \sim C(a,h)$, where $a = a_1 + \cdots + a_n$ and $h = h_1 + \cdots + h_n$.

Exercise 12.1 Let $Y_k \sim C(a,h)$ $(k = 1, 2, \ldots, n)$ be independent random variables, and $Y = (Y_1 + \cdots + Y_n)/n$. Show then that Y also has the Cauchy $C(a,h)$ distribution.

Exercise 12.2 Let $Y_k \sim C(a,h)$ $(k = 1, 2, \ldots, n)$ be independent random variables. Find the distribution of a linear combination $Y = c_1 Y_1 + \cdots + c_n Y_n$.

12.5   Decompositions

The result in (12.9) implies that any Cauchy distribution can be represented as a sum of two or more random variables also having the Cauchy distributions. Moreover, for any $n = 1, 2, \ldots$, the Cauchy $C(a, h)$ distribution is expressed as a sum of n independent random variables, each of them having the Cauchy $C(a/n, h/n)$ distribution (see Exercise 12.1). This means that any Cauchy distribution is infinitely divisible.

12.6   Stable Distributions

A characteristic function $f(t)$ and the corresponding distribution are said to be *stable* if for any $a_1 > 0$ and $a_2 > 0$, there exist constants $a > 0$ and b such that

$$f(a_1 t) f(a_2 t) = e^{itb} f(at). \tag{12.10}$$

Exercise 12.3 Show that any Cauchy distribution is stable.

12.7   Transformations

Let $X \sim C(0, 1)$ and $Y = 1/X$. It turns out that Y also has the standard Cauchy distribution. In fact, the pdf's of X and Y satisfy the relationship

$$x^2 p_Y(x) = p_X(1/x). \tag{12.11}$$

From (12.11), we simply obtain

$$p_Y(x) = \frac{1}{\pi(1 + x^{-2})x^2} = \frac{1}{\pi(1 + x^2)}, \qquad -\infty < x < \infty.$$

Remark 12.1 The fact that

$$X \stackrel{d}{=} 1/X \tag{12.12}$$

does not, however, imply that X must have the Cauchy $C(0, 1)$ distribution. To see this, let us take V to be any symmetric random variable, and $X = e^V$. Then, since

$$V \stackrel{d}{=} -V,$$

we readily have

$$X = e^V \stackrel{d}{=} e^{-V} = 1/X.$$

Exercise 12.4 Use (12.5) to show that if $X \sim C(0,1)$, then $V = \frac{1}{2}\left(X - \frac{1}{X}\right)$ also has the Cauchy $C(0,1)$ distribution.

Exercise 12.5 Let $X \sim C(a,h)$ and $Y = 1/X$. Show then that

$$Y \sim C\left(\frac{a}{a^2 + h^2}, \frac{h}{a^2 + h^2}\right).$$

CHAPTER 13

TRIANGULAR DISTRIBUTION

13.1 Introduction

In Chapter 11 we derived the pdf of the sum of two independent standard uniform $U(0,1)$ random variables as

$$p_X(x) = \begin{cases} x & \text{if } 0 \leq x \leq 1 \\ 2 - x & \text{if } 1 \leq x \leq 2. \end{cases} \qquad (13.1)$$

Because of the triangular shape of $p(x)$, the corresponding distribution is called as a *triangular distribution*.

13.2 Notations

The standard triangular distribution has its pdf as

$$p_X(x) = \begin{cases} \dfrac{2x}{p} & \text{if } 0 \leq x \leq p \\ \dfrac{2(1-x)}{1-p} & \text{if } p \leq x \leq 1, \end{cases} \qquad (13.2)$$

where $0 < p < 1$. We will denote the random variable X which has its pdf as in (13.2) by

$$X \sim \text{Tr}(p, 0, 1).$$

The linear transformation $Y = a + hX$ then provides a general form of triangular density functions, since

$$p_Y(x) = \frac{1}{h}\, p_X\left(\frac{x-a}{h}\right) \qquad \text{if } a \leq x \leq a+h.$$

We denote this general triangular random variable by

$$Y \sim \text{Tr}(p, a, h), \qquad 0 < p < 1, \ -\infty < a < \infty, \ h > 0,$$

which has the pdf

$$p_Y(x) = \begin{cases} \dfrac{2(x-a)}{h^2 p} & \text{if } a \le x \le a + ph \\[2mm] \dfrac{2(h+a-x)}{h^2(1-p)} & \text{if } a + ph \le x \le a + h. \end{cases} \tag{13.3}$$

If $p = \frac{1}{2}$, the distribution of Y is called the *symmetrical triangular distribution* (symmetric around the point $a + h/2$).

13.3 Moments

Let $X \sim \text{Tr}(p, 0, 1)$. All moments of X exist since it is a bounded random variable.

Moments about zero:

$$\begin{aligned} \alpha_n &= EX^n = \int_0^p \frac{2}{p}\, x^{n+1}\, dx + \int_p^1 \frac{2}{1-p}\, x^n (1-x)\, dx \\[2mm] &= \frac{2}{n+2}\, p^{n+1} + \frac{2}{1-p} \left(\frac{1 - p^{n+1}}{n+1} - \frac{1 - p^{n+2}}{n+2} \right) \\[2mm] &= \frac{2(1 + p + p^2 + \cdots + p^n)}{(n+1)(n+2)}, \qquad n = 1, 2, \ldots. \end{aligned} \tag{13.4}$$

In particular, we have the following expressions for the first four moments:

$$\alpha_1 = EX = \frac{1+p}{3}, \tag{13.5}$$

$$\alpha_2 = EX^2 = \frac{1 + p + p^2}{6}, \tag{13.6}$$

$$\alpha_3 = EX^3 = \frac{1 + p + p^2 + p^3}{10}, \tag{13.7}$$

$$\alpha_4 = EX^4 = \frac{1 + p + p^2 + p^3 + p^4}{15}. \tag{13.8}$$

Central moments:

From (13.5) and (13.6), we readily find the variance of X to be

$$\beta_2 = \text{Var}\, X = \alpha_2 - \alpha_1^2 = \frac{1 - p + p^2}{18}. \tag{13.9}$$

Similarly, from (13.5)–(13.8), we find the third and fourth central moments of X to be

$$\begin{aligned} \beta_3 &= \alpha_3 - 3\alpha_2\alpha_1 + 2\alpha_1^3 \\[2mm] &= \frac{2 - 3p - 3p^2 + 2p^3}{270}, \end{aligned} \tag{13.10}$$

$$\begin{aligned} \beta_4 &= \alpha_4 - 4\alpha_3\alpha_1 + 6\alpha_2\alpha_1^2 - 3\alpha_1^4 \\[2mm] &= \frac{1 - 2p + 3p^2 - 2p^3 + p^4}{135}, \end{aligned} \tag{13.11}$$

respectively.

Moments about p:

For the triangular distributions, due to the form of the pdf in (13.2), it is easier to find the moments about p rather than the central moments [moments about the mean $EX = (1+p)/3$]. The nth moment about p is obtained from (13.2) to be

$$E(X - p)^n = \frac{2\left\{(-1)^n p^{n+1} + (1-p)^{n+1}\right\}}{(n+1)(n+2)}. \tag{13.12}$$

Shape characteristics:

From (13.9) and (13.10) we obtain the coefficient of skewness as

$$\gamma_1 = \frac{\beta_3}{\beta_2^{3/2}} = \frac{\sqrt{2}(1-2p)(1+p)(2-p)}{5(1-p+p^2)^{3/2}}. \tag{13.13}$$

It is then clear from (13.13) that the distribution is positively skewed for $p < \frac{1}{2}$ and is negatively skewed for $p > \frac{1}{2}$. The coefficient of skewness is 0 when $p = \frac{1}{2}$ (in this case, the distribution is in fact symmetric).

From (13.9) and (13.11) we obtain the coefficient of kurtosis as

$$\gamma_2 = \frac{\beta_4}{\beta_2^2} = \frac{12(1 - 2p + 3p^2 - 2p^3 + p^4)}{5(1 - p + p^2)^2} = \frac{12}{5} = 2.4, \tag{13.14}$$

which reveals that the distribution is platykurtic for all choices of p.

13.4 Characteristic Function

Let $X \sim \mathrm{Tr}(p, 0, 1)$. Then, the characteristic function $f_X(t)$ has the form

$$f_X(t) = Ee^{itX} = \frac{2}{p(1-p)t^2}\left\{e^{itp} - p\, e^{it} - (1-p)\right\}. \tag{13.15}$$

In the general case when $X \sim \mathrm{Tr}(p, a, h)$, the corresponding characteristic function is given by

$$f_X(t) = \frac{2e^{ita}}{p(1-p)h^2t^2}\left\{e^{itph} - p\, e^{ith} - (1-p)\right\}. \tag{13.16}$$

The most interesting case is $Y \sim \mathrm{Tr}\left(\frac{1}{2}, -1, 2\right)$, when the pdf of Y is

$$p_Y(x) = \max\{0, 1 - |x|\}, \tag{13.17}$$

and the characteristic function of Y is

$$f_Y(t) = \frac{2 - e^{it} - e^{-it}}{t^2} = \left(\frac{2\sin(t/2)}{t}\right)^2 = \frac{4\sin^2(t/2)}{t^2}. \tag{13.18}$$

Note that $f(t) = (2/t)\sin(t/2)$ is the characteristic function corresponding to the symmetric uniform $U\left(-\frac{1}{2}, \frac{1}{2}\right)$ distribution and, hence, the triangular $\mathrm{Tr}\left(\frac{1}{2}, -1, 2\right)$ distribution is the convolution of two uniform distributions concentrated on the same interval $\left(-\frac{1}{2}, \frac{1}{2}\right)$.

Consider the following expression, which relates the pdf and the characteristic function of a certain random variable Y:

$$p_Y(t) = \frac{1}{2\pi} \int_{-\infty}^{\infty} e^{-itx} f_Y(x) \, dx. \qquad (13.19)$$

In the case when $Y \sim \mathrm{Tr}\left(\frac{1}{2}, -1, 2\right)$, (13.19) gives

$$\max\{0, 1 - |t|\} = \frac{1}{2\pi} \int_{-\infty}^{\infty} e^{-itx} \left(\frac{4\sin^2(x/2)}{x^2}\right) dx \qquad (13.20)$$

which is equivalent to

$$\max\{0, 1 - |t|\} = \int_{-\infty}^{\infty} e^{itx} \left(\frac{2\sin^2(x/2)}{\pi}\right) dx. \qquad (13.21)$$

Setting $t = 0$ in (13.21), we simply obtain

$$\int_{-\infty}^{\infty} \left(\frac{2\sin^2(x/2)}{\pi}\right) dx = 1, \qquad (13.22)$$

and, consequently, the nonnegative function $p(x) = (2/\pi)\sin^2(x/2)$ is the pdf of some random variable V. We can, .therefore, see from (13.21) that $f(t) = \max\{0, 1 - |t|\}$ is indeed the characteristic function corresponding to the pdf $p(x) = (2/\pi)\sin^2(x/2)$. Hence, for any $A > 0$, the function $f_A(t) = f(At) = \max\{0, 1 - A|t|\}$ is also a characteristic function. Characteristic functions $f_A(t)$ are used to prove the *Pólya criterion*, giving a sufficient condition for a real function to be a characteristic function:

If a real even continuous function $g(t)$ [with $g(0) = 1$] is convex on $(0, \infty)$ and $\lim_{t\to\infty} g(t) = p$, $0 \le p \le 1$, then $g(t)$ is a characteristic function of some random variable.

CHAPTER 14

POWER DISTRIBUTION

14.1 Introduction

In Chapter 11 [Eq. (11.32)] we found that the distribution of maximal value $M(n) = \max\{U_1, \ldots, U_n\}$ of n independent uniform $U(0,1)$ random variables $U_1, \ldots, U_n$ has the following cdf:

$$H_n(x) = x^n, \qquad 0 < x < 1. \tag{14.1}$$

Note that if $U \sim U(0,1)$ and $X = U^{1/n}$, then also X has cdf (14.1), which is a special case of the *power distribution*.

14.2 Notations

The standard power distribution function has the form

$$F_\alpha(x) = x^\alpha, \qquad 0 < x < 1, \ \alpha > 0, \tag{14.2}$$

and the corresponding pdf is

$$p_\alpha(x) = \alpha x^{\alpha-1}, \qquad 0 < x < 1, \ \alpha > 0. \tag{14.3}$$

In the special case when $\alpha = 1$, we have the standard uniform distribution.

The linear transformation gives us a general form of the power distribution. The corresponding cdf is

$$G_\alpha(x) = \left(\frac{x-a}{h}\right)^\alpha, \qquad a < x < a+h, \ \alpha > 0, \tag{14.4}$$

and the pdf is

$$g_\alpha(x) = \alpha h^{-\alpha}(x-a)^{\alpha-1}, \tag{14.5}$$

where $\alpha > 0$, $-\infty < a < \infty$ and $h > 0$ are the shape, location, and scale parameters of the power distribution, respectively. We will use the notation

$$X \sim Po(\alpha, a, h)$$

to denote the random variable X having cdf (14.4) and pdf (14.5).

Remark 14.1 Let U have the standard uniform $U(0,1)$ distribution, and $X \sim Po(\alpha, a, h)$. Then,

$$X \stackrel{d}{=} a + hU^{1/\alpha}. \tag{14.6}$$

14.3 Distributions of Maximal Values

Let $X_1, X_2, \ldots, X_n$ be independent random variables having the same power $Po(\alpha, a, h)$ distribution, and $M_n = \max\{X_1, X_2, \ldots, X_n\}$. It is then easy to see that

$$P\{M_n \le x\} = \left(\frac{x - a}{h}\right)^{\alpha n}, \qquad a < x < a + h,$$

i.e.,

$$M_n \sim Po(\alpha n, a, h).$$

Thus, the power distributions are *closed under maxima*.

Exercise 14.1 Let $X_1, X_2, \ldots, X_n$ be a random sample from the standard power $Po(\alpha, 0, 1)$ distribution, and let $X_{1,n} < X_{2,n} < \cdots < X_{n,n}$ be the corresponding order statistics. Then, from the pdf of $X_{k,n}$ $(k = 1, 2 \ldots, n)$ given by

$$p_{k,n}(x) = \frac{n!}{(k - 1)! \, (n - k)!} \{F_\alpha(x)\}^{k-1} \{1 - F_\alpha(x)\}^{n-k} \, p_\alpha(x), \ 0 < x < 1,$$

where $F_\alpha(x)$ and $p_\alpha(x)$ are as in (14.2) and (14.3), derive the following formulas:

$$EX_{k,n} = \frac{n! \, \Gamma\left(k + 1/\alpha\right)}{(k - 1)! \, \Gamma\left(n + 1 + 1/\alpha\right)}$$

and

$$E\left(X_{k,n}^2\right) = \frac{n! \, \Gamma\left(k + 2/\alpha\right)}{(k - 1)! \, \Gamma\left(n + 1 + 2/\alpha\right)}.$$

Exercise 14.2 Let $X_1, X_2, \ldots, X_n$ be a random sample from the standard power $Po(\alpha, 0, 1)$ distribution, and let $X_{1,n} < X_{2,n} < \cdots < X_{n,n}$ be the corresponding order statistics. Further, let

$$W_1 = \frac{X_{1,n}}{X_{2,n}}, \ W_2 = \frac{X_{2,n}}{X_{3,n}}, \ \cdots \ , W_{n-1} = \frac{X_{n-1,n}}{X_{n,n}}, \ W_n = X_{n,n}.$$

Prove that $W_1, W_2, \ldots, W_n$ are independent random variables with W_k distributed as $Po(k\alpha, 0, 1)$.

Exercise 14.3 By making use of the distributional result above, derive the expressions for $EX_{k,n}$ and $E(X_{k,n}^2)$ presented in Exercise 14.1.

14.4 Moments

The relation in (14.6) enables us to express moments of the power distribution in terms of moments of the uniform distribution.

Moments about zero:
  Let $X \sim Po(\alpha, a, h)$. Then,

$$
\alpha_n = EX^n = E(a + hU^{1/\alpha})^n \;=\; \sum_{k=0}^{n} \binom{n}{k} h^k a^{n-k} EU^{k/\alpha}
$$

$$
= \sum_{k=0}^{n} \frac{\alpha}{\alpha+k} \binom{n}{k} h^k a^{n-k}. \qquad (14.7)
$$

In particular, we have

$$
\alpha_1 = EX = a + \frac{h\alpha}{\alpha+1} \qquad (14.8)
$$

and

$$
\alpha_2 = EX^2 = a^2 + \frac{2ah\alpha}{\alpha+1} + \frac{h^2\alpha}{\alpha+2}. \qquad (14.9)
$$

From (14.8) and (14.9), we also obtain the variance of X as

$$
\beta_2 = E(X - EX)^2 = \alpha_2 - \alpha_1^2 = \frac{h^2\alpha}{(\alpha+1)^2(\alpha+2)}. \qquad (14.10)
$$

Plots of power density function are presented in Figure 14.1 for some choices of α.

Exercise 14.4 From (14.7), derive the expressions for the coefficients of skewness and kurtosis (γ_1 and γ_2).

Exercise 14.5 From the expression of γ_1 derived in Exercise 14.1, discuss the nature of the skewness of the power distribution.

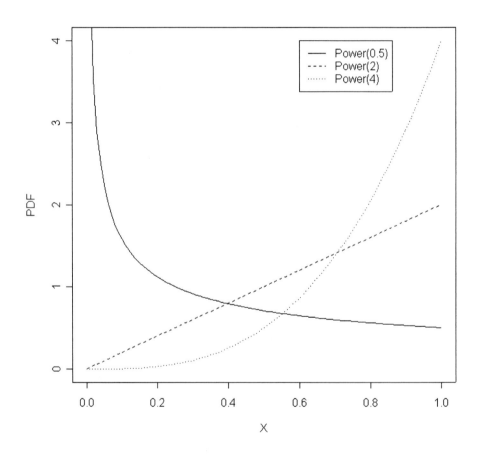

Figure 14.1. Plots of power density function

14.5 Entropy

Recall from Section 1.5 that the entropy $H(X)$ of a random variable having a pdf $p(x)$ is defined as

$$H(X) = -\int_D p(x) \log p(x) \, dx,$$

where

$$D = \{x : p(x) > 0\}.$$

Exercise 14.6 For $X \sim Po(\alpha, a, h)$, show that

$$H(X) = \log h - \log \alpha + 1 - \frac{1}{\alpha} . \tag{14.11}$$

14.6 Characteristic Function

Let X follow the standard power distribution $Po(\alpha, 0, 1)$. Then, the corresponding characteristic function has the form

$$
\begin{aligned}
f_\alpha(t) &= \alpha \int_0^1 e^{itx} x^{\alpha-1} \, dx = \alpha \sum_{k=0}^{\infty} \int_0^1 \frac{(itx)^k x^{\alpha-1}}{k!} \, dx \\
&= \alpha \sum_{k=0}^{\infty} \frac{(it)^k}{k! \, (k+\alpha)} .
\end{aligned}
\tag{14.12}
$$

The RHS of (14.12) can be simplified for integer α. For example,

$$f_1(t) = \int_0^1 e^{itx} \, dx = \frac{e^{it} - 1}{it} \tag{14.13}$$

is the characteristic function of the standard uniform $U(0, 1)$ random variable. It is also easy to check that the function $f_n(t)$ satisfies the following recurrence relation:

$$f_n(t) = \frac{n \left\{ e^{it} - f_{n-1}(t) \right\}}{it}, \qquad n = 2, 3, \dots . \tag{14.14}$$

Exercise 14.7 Prove (14.14) and then show that for any $n = 1, 2, \ldots$ the following expression for $f_n(t)$ is true:

$$f_n(t) = (-1)^{n+1} n! \, (it)^{-n} e^{it} \sum_{k=0}^{n-1} \frac{(-it)^k}{k!} + (-1)^n n! \, (it)^{-n}. \qquad (14.15)$$

In particular, deduce that

$$f_2(t) = \frac{2e^{it}(1 - it - e^{-it})}{t^2} \qquad (14.16)$$

and

$$f_3(t) = \frac{6e^{it}\left\{1 - it - t^2/2 - e^{-it}\right\}}{(it)^3}. \qquad (14.17)$$

Since we know the characteristic function of the standard power distribution, it will be possible to write the characteristic function of the general power distribution. In fact, the relation

$$Y \stackrel{d}{=} a + hX,$$

where $X \sim Po(\alpha, 0, 1)$ and $Y \sim Po(\alpha, a, h)$, readily implies that

$$f_Y(t) = Ee^{itY} = e^{iat} f_\alpha(ht),$$

with $f_\alpha(t) = Ee^{itX}$ as given in (14.12) for any $\alpha > 0$, and as in (14.15) for integer values of α.

CHAPTER 15

PARETO DISTRIBUTION

15.1   Introduction

The standard power distribution $Po(\alpha, 0, 1)$, $\alpha > 0$, which we discussed in Chapter 14, was noted earlier [see (14.6)] to be the same as the distribution of $U^{1/\alpha}$ (for $\alpha > 0$), where U is the standard uniform $U(0, 1)$ random variable. Let us now consider the distribution of $U^{1/\alpha}$ for negative values of α. Let

$$X \stackrel{d}{=} U^{-1/\alpha}, \qquad \alpha > 0.$$

We can then see that the corresponding cdf and pdf are given by

$$F_\alpha(x) = P\{U^{-1/\alpha} < x\} = P\{U > x^{-\alpha}\} = 1 - x^{-\alpha}, \quad x \geq 1, \qquad (15.1)$$

and

$$p_\alpha(x) = \begin{cases} \alpha x^{-(\alpha+1)} & \text{if } x \geq 1 \\ 0 & \text{if } x < 1. \end{cases} \qquad (15.2)$$

A book-length account of Pareto distributions, discussing in great detail their various properties and applications, is available [Arnold (1983)].

15.2   Notations

A random variable X with cdf (15.1) and pdf (15.2) is said to have the *standard Pareto distribution*. A linear transformation

$$Y = a + h\xi$$

gives a general form of Pareto distributions with pdf

$$p_Y(x) = \begin{cases} \dfrac{\alpha h^\alpha}{(x-a)^{\alpha+1}} & \text{if } x \geq a + h \\ 0 & \text{if } x < a + h, \end{cases} \qquad (15.3)$$

and cdf

$$F_Y(x) = 1 - \left(\frac{h}{x-a}\right)^\alpha, \qquad x \geq a + h, \ -\infty < a < \infty, \ h > 0. \qquad (15.4)$$

We will use the notation

$$Y \sim Pa(\alpha, a, h), \qquad \alpha > 0, \ -\infty < a < \infty, \ h > 0,$$

to denote the random variable Y having the Pareto distribution with pdf (15.3) and cdf (15.4). Then,

$$X \sim Pa(\alpha, 0, 1), \qquad \alpha > 0,$$

corresponds to the standard Pareto distribution with pdf (15.2) and cdf (15.1). Note that if

$$V \sim Po(\alpha, 0, 1),$$

then

$$X = \frac{1}{V} \sim Pa(\alpha, 0, 1).$$

It is easy to see that if $Y \sim Pa(\alpha, a, h)$, then the following representation of Y in terms of the standard uniform random variable U is valid:

$$Y \stackrel{d}{=} a + hU^{-1/\alpha}. \tag{15.5}$$

Exercise 15.1 Prove the relation in (15.5).

15.3   Distributions of Minimal Values

Just as the power distributions have simple form for the distributions of maximal values $M_n = \max\{X_1, X_2, \ldots, X_n\}$, the Pareto distributions possess convenient expressions for minimal values. Let $X_1, X_2, \ldots, X_n$ be independent random variables, and let

$$X_k \sim Pa(\alpha_k, a, h), \qquad k = 1, 2, \ldots, n,$$

i.e., these random variables have the same location and scale parameters but have different values of the shape parameter α_k. Now, let

$$m_n = \min\{X_1, X_2, \ldots, X_n\}.$$

It is then easy to see that

$$P\{m_n \le x\} = 1 - P\{X_1 > x\} \cdots P\{X_n > x\} \quad = \quad \frac{h^{\alpha(n)}}{(x-a)^{\alpha(n)}},$$
$$x \ge a + h, \quad (15.6)$$

where

$$\alpha(n) = \alpha_1 + \alpha_2 + \cdots + \alpha_n.$$

We simply note from (15.6) that

$$m_n \sim Pa(\alpha(n), a, h).$$

Thus, the Pareto distributions are *closed under minima*.

Exercise 15.2 Let $X_1, X_2, \ldots, X_n$ be a random sample from the standard Pareto $Pa(\alpha, 0, 1)$ distribution, and let $X_{1,n} < X_{2,n} < \cdots < X_{n,n}$ be the corresponding order statistics. Then, from the pdf of $X_{k,n}$ $(k = 1, 2 \ldots, n)$ given by

$$p_{k,n}(x) = \frac{n!}{(k-1)!\,(n-k)!} \{F_\alpha(x)\}^{k-1} \{1 - F_\alpha(x)\}^{n-k} p_\alpha(x), \ 0 < x < 1,$$

where $F_\alpha(x)$ and $p_\alpha(x)$ are as in (15.1) and (15.2), derive the following formulas:

$$EX_{k,n} = \frac{n!\,\Gamma\,(n-k+1-1/\alpha)}{(n-k)!\,\Gamma\,(n+1-1/\alpha)}$$

and

$$E\left(X_{k,n}^2\right) = \frac{n!\,\Gamma\,(n-k+1-2/\alpha)}{(n-k)!\,\Gamma\,(n+1-2/\alpha)}\,.$$

Exercise 15.3 Let $X_1, X_2, \ldots, X_n$ be a random sample from the standard pareto $Pa(\alpha, 0, 1)$ distribution, and let $X_{1,n} < X_{2,n} < \cdots < X_{n,n}$ be the corresponding order statistics. Further, let

$$W_1 = X_{1,n}, \ W_2 = \frac{X_{2,n}}{X_{1,n}}, \ W_3 = \frac{X_{3,n}}{X_{2,n}}, \ \ldots, W_n = \frac{X_{n,n}}{X_{n-1,n}}.$$

Prove that $W_1, W_2, \ldots, W_n$ are independent random variables with W_k distributed as $Pa((n-k+1)\alpha, 0, 1)$.

Exercise 15.4 By making use of the distributional result above, derive the expressions for $EX_{k,n}$ and $E(X_{k,n}^2)$ presented in Exercise 15.2.

Exercise 15.5 Let $V_1, V_2, \ldots, V_n$ be a random sample from the standard power $Po(\alpha, 0, 1)$ distribution, and let $V_{1,n} < V_{2,n}, \ldots, V_{n,n}$ be the corresponding order statistics. Further, let $X_1, X_2, \ldots, X_n$ be a random sample from the standard Pareto $Pa(\alpha, 0, 1)$ distribution, and let $X_{1,n} < X_{2,n} < \cdots < X_{n,n}$ be the corresponding order statistics. Then, show that

$$X_{k,n} \stackrel{d}{=} \frac{1}{V_{n-k+1,n}} \qquad \text{for } k = 1, 2, \ldots, n.$$

Exercise 15.6 By using the propoerty in Exercise 15.5 along with the distributional result for order statistics from power function distribution presented before in Exercise 14.2, establish the result in Exercise 15.3.

15.4 Moments

Let $X \sim Pa(\alpha, 0, 1)$ and $Y = a + hX \sim Pa(\alpha, a, h)$. Unlike the power distributions which have bounded support and hence possess finite moments of all orders, the Pareto distribution $Pa(\alpha, a, h)$ takes on values in an infinite interval $(a + h, \infty)$ and its moments $E\eta^\beta$ and central moments $E(Y - EY)^\beta$ are finite only when $\beta < \alpha$.

Moments about zero:

If $X \sim Pa(\alpha, 0, 1)$, then we know that X can be expressed in terms of the standard uniform random variable U as

$$X \stackrel{d}{=} U^{-1/\alpha}.$$

Hence,

$$
\begin{aligned}
\alpha_n &= EX^n = EU^{-n/\alpha} \\
&= \int_0^1 x^{-n/\alpha}\, dx = \frac{\alpha}{\alpha - n} \qquad \text{if } n < \alpha,
\end{aligned}
\tag{15.7}
$$

and

$$\alpha_n = \infty \qquad \text{if } n \ge \alpha.$$

In the general case when $Y \sim Pa(\alpha, a, h)$, the relation in (15.5) holds and consequently, we obtain

$$
\begin{aligned}
\alpha_n &= EY^n = E(a + hU^{-1/\alpha})^n \\
&= \sum_{k=0}^{n} \binom{n}{k} h^k a^{n-k} EU^{-k/\alpha} \\
&= \sum_{k=0}^{n} \frac{\alpha}{\alpha - k} \binom{n}{k} h^k a^{n-k} \qquad \text{if } n > \alpha
\end{aligned}
\tag{15.8}
$$

and $\alpha_n = \infty$ for $n \ge \alpha$. In particular, we have

$$\alpha_1 = EY = a + \frac{h\alpha}{\alpha - 1} \qquad \text{if } \alpha > 1 \tag{15.9}$$

and

$$\alpha_2 = EY^2 = a^2 + \frac{2ah\alpha}{\alpha - 1} + \frac{h^2\alpha}{\alpha - 2} \qquad \text{if } \alpha > 2. \tag{15.10}$$

Central moments:

The relation in (15.5) can also be used to find the central moments of a Pareto distribution, as follows:

$$
\begin{aligned}
\beta_n &= E(Y - EY)^n = h^n E(U^{-1/\alpha} - EU^{-1/\alpha})^n \\
&= h^n E\left(U^{-1/\alpha} - \frac{\alpha}{\alpha - 1}\right)^n \\
&= h^n \sum_{k=0}^{n} \binom{n}{k} (-1)^{n-k} \alpha^{n-k} (\alpha - 1)^{k-n} EU^{-k/\alpha} \\
&= h^n \sum_{k=0}^{n} \binom{n}{k} (-1)^{n-k} \alpha^{n-k+1} \frac{(\alpha - 1)^{k-n}}{\alpha - k}
\end{aligned}
\tag{15.11}
$$

for $n < \alpha$.

From (15.11) we find the variance of Y to be

$$
\begin{aligned}
\mathrm{Var}\, Y = \beta_2 &= h^2 \left(\frac{\alpha}{\alpha - 2} - \frac{\alpha^2}{(\alpha - 1)^2}\right) \\
&= \frac{\alpha h^2}{(\alpha - 2)(\alpha - 1)^2} \qquad \text{if } \alpha > 2.
\end{aligned}
\tag{15.12}
$$

Plots of Pareto density function are presented in Figure 15.1 for some choices of α.

15.5 Entropy

The entropy $H(Y)$ of $Y \sim Pa(\alpha, a, h)$ is defined by

$$
H(Y) = -\int_{a+h}^{\infty} p(x) \log p(x)\, dx,
$$

where $p(x)$ is as given in (15.3).

Exercise 15.7 If $Y \sim Pa(\alpha, a, h)$, show that

$$
H(Y) = -(1 + 2\alpha) \log h - \log \alpha + 1 + \frac{1}{\alpha}.
\tag{15.13}
$$

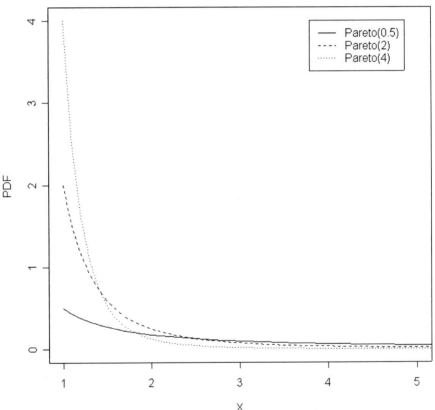

Figure 15.1. Plots of Pareto density function

CHAPTER 16

BETA DISTRIBUTION

16.1 Introduction

Let a stick of length 1 be broken at random into $(n + 1)$ pieces. In other words, this means that n breaking points $U_1, \ldots, U_n$ of the unit interval are taken independently from the uniform $U(0, 1)$ distribution. Arranging these random coordinates $U_1, \ldots, U_n$ in increasing order of magnitude, we obtain the *uniform order statistics*

$$0 \leq U_{1,n} \leq \cdots \leq U_{n,n} \leq 1$$

[see (11.39)], where, for instance,

$$U_{1,n} = \min\{U_1, \ldots, U_n\}$$

and

$$U_{n,n} = \max\{U_1, \ldots, U_n\}.$$

The cdf $F_{k,n}(x)$ of $U_{k,n}$ was obtained earlier in (11.41) as

$$F_{k,n}(x) = \frac{n!}{(k-1)!\,(n-k)!} \int_0^x t^{k-1}(1-t)^{n-k}\,dt = I_x(k, n-k+1),$$

where

$$I_x(\alpha, \beta) = \frac{1}{B(\alpha, \beta)} \int_0^x t^{\alpha-1}(1-t)^{\beta-1}\,dt \qquad (16.1)$$

denotes the incomplete beta function, and

$$B(\alpha, \beta) = \int_0^1 t^{\alpha-1}(1-t)^{\beta-1}\,dt = \frac{\Gamma(\alpha)\Gamma(\beta)}{\Gamma(\alpha+\beta)} \qquad (16.2)$$

is the complete beta function. The corresponding pdf $f_{k,n}(x)$ of $U_{k,n}$ has the form

$$f_{k,n}(x) = \frac{n!}{(k-1)!\,(n-k)!}\,x^{k-1}(1-x)^{n-k}, \qquad 0 < x < 1. \qquad (16.3)$$

16.2   Notations

We say that the order statistic $U_{k,n}$ has the beta distribution with shape parameters k and $n - k + 1$. More generally, we say that a random variable X has the *standard beta distribution* with parameters $p > 0$ and $q > 0$ if its pdf is given by

$$
\begin{aligned}
p_X(x) &= \frac{1}{B(p,q)}\, x^{p-1}(1-x)^{q-1} \\
&= \frac{\Gamma(p+q)}{\Gamma(p)\Gamma(q)}\, x^{p-1}(1-x)^{q-1}, \qquad 0 < x < 1. \quad (16.4)
\end{aligned}
$$

A linear transformation

$$
Y = a + hX, \qquad -\infty < a < \infty,\ h > 0
$$

gives us the general form of beta distributions with pdf

$$
\begin{aligned}
p_Y(x) &= \frac{1}{h^{p+q-1}B(p,q)}\, (x-a)^{p-1}(h+a-x)^{q-1} \\
&= \frac{\Gamma(p+q)}{h^{p+q-1}\Gamma(p)\Gamma(q)}\, (x-a)^{p-1}(h+a-x)^{q-1}, \\
&\qquad\qquad\qquad\qquad\qquad a < x < a + h. \quad (16.5)
\end{aligned}
$$

The random variable with pdf (16.5) is denoted by

$$
Y \sim \mathrm{beta}(p,q,a,h).
$$

For the random variable $X \sim \mathrm{beta}(p,q,0,1)$ having the standard beta distribution with pdf (16.4), we will use the simplified notation

$$
X \sim \mathrm{beta}(p,q).
$$

The most important special cases of beta distributions are the uniform ($p = 1$, $q = 1$) and power ($p > 0$, $q = 1$) distributions. We also make a note here that the special case of beta distributions with $p = q = \frac{1}{2}$ is known as the *arcsine distribution* (discussed in Chapter 17), and the case $q = 1 - p$, $0 < p < 1$, corresponds to the *generalized arcsine distribution*. The uniform and arcsine distributions are symmetric (about their expectations) distributions. The same property also holds for all $\mathrm{beta}(p,p,a,h)$ distributions (that is, with equal parameters p and q), and they are symmetric about their expectation $a + h/2$.

16.3   Mode

If $p > 1$ and $q > 1$, then the $\mathrm{beta}(p,q)$ distribution is unimodal, and its mode (the point at which the density function $p_X(x)$ takes its maximal value) is at $x = \dfrac{p-1}{p+q-2}$, and the density function in this case is a bell-shaped curve. If

$p < 1$ (or $q < 1$), then $p_X(x)$ tends to infinity when $x \to 0$ (or when $x \to 1$). If $p < 1$ and $q < 1$, then $p_X(x)$ has a U-shaped form and it tends to infinity when $x \to 0$ as well as when $x \to 1$. If $p = 1$ and $q = 1$, then $p_X(x) = 1$ for all $0 < x < 1$ (which, as noted earlier, is the standard uniform density).

Plots of beta density function are presented in Figures 16.1–16.3 for different choices of p and q.

16.4   Some Transformations

It is quite easy to see that if $X \sim \text{beta}(p, q)$, then $1 - X \sim \text{beta}(q, p)$. Now, let us find the distribution of the random variable $V = 1/X$.

Exercise 16.1 Show that

$$p_V(x) = \frac{1}{B(p, q)} \frac{(x - 1)^{q-1}}{x^{p+q}} \,, \qquad x > 1. \tag{16.6}$$

Taking $q = 1$ in (16.6), we obtain the pdf of Pareto $\text{Pa}(p, 0, 1)$ distribution. The density function $p_W(x)$ of the random variable

$$W = V - 1 = \frac{1}{X} - 1 = \frac{1 - X}{X}$$

takes on the form

$$p_W(x) = \frac{1}{B(p, q)} \frac{x^{q-1}}{(1 + x)^{p+q}} \,, \qquad x > 0. \tag{16.7}$$

Distribution with pdf (16.7) is sometimes called the *beta distribution of the second kind*.

16.5   Moments

Since the beta distribution has bounded support, all its moments exist. Consider the standard beta distributed random variable $X \sim \text{beta}(p, q)$. Since $0 \leq X \leq 1$, we can conclude that

$$0 \leq EX^\alpha \leq 1$$

and

$$E|X - EX|^\alpha \leq 1$$

for any $\alpha \geq 0$.

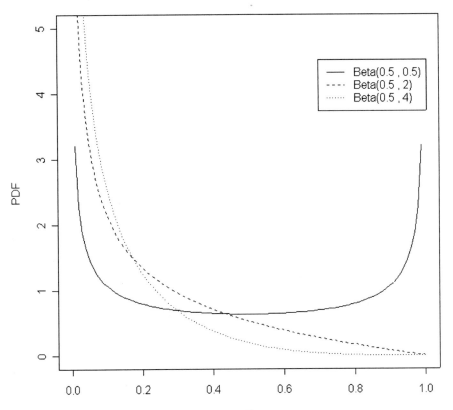

Figure 16.1. Plots of beta density function when $p = 0.5$

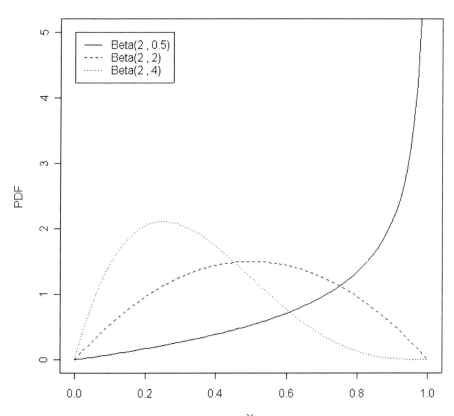

Figure 16.2. Plots of beta density function when $p = 2.0$

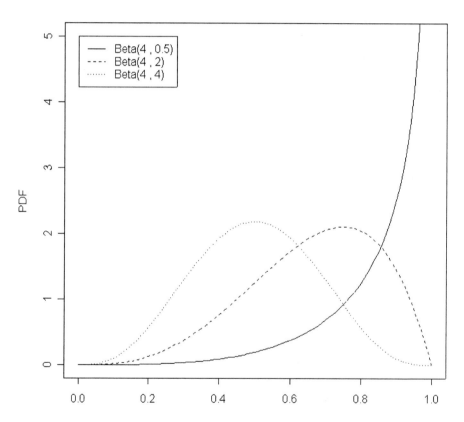

Figure 16.3. Plots of beta density function when $p = 4.0$

Moments about zero:

Let $X \sim \text{beta}(p, q)$. Then

$$
\begin{aligned}
\alpha_n \;=\; EX^n &= \frac{1}{B(p, q)} \int_0^1 x^{n+p-1}(1 - x)^{q-1} \, dx \\
&= \frac{B(n + p, q)}{B(p, q)} \\
&= \frac{\Gamma(n + p)\Gamma(q)\Gamma(p + q)}{\Gamma(n + p + q)\Gamma(p)\Gamma(q)} \\
&= \frac{\Gamma(n + p)\Gamma(p + q)}{\Gamma(n + p + q)\Gamma(p)} \quad\quad (16.8)
\end{aligned}
$$

and, in particular,

$$
\alpha_1 = EX \;=\; \frac{\Gamma(p + 1)\Gamma(p + q)}{\Gamma(p + q + 1)\Gamma(p)} = \frac{p}{p + q} \;, \quad\quad (16.9)
$$

$$
\alpha_2 = EX^2 \;=\; \frac{\Gamma(p + 2)\Gamma(p + q)}{\Gamma(p + q + 2)\Gamma(p)} = \frac{p(p + 1)}{(p + q)(p + q + 1)} \;, \quad\quad (16.10)
$$

$$
\begin{aligned}
\alpha_3 = EX^3 &= \frac{\Gamma(p + 3)\Gamma(p + q)}{\Gamma(p + q + 3)\Gamma(p)} \\
&= \frac{p(p + 1)(p + 2)}{(p + q)(p + q + 1)(p + q + 2)} \;, \quad\quad (16.11)
\end{aligned}
$$

$$
\begin{aligned}
\alpha_4 = EX^4 &= \frac{\Gamma(p + 4)\Gamma(p + q)}{\Gamma(p + q + 4)\Gamma(p)} \\
&= \frac{p(p + 1)(p + 2)(p + 3)}{(p + q)(p + q + 1)(p + q + 2)(p + q + 3)} \;. \quad\quad (16.12)
\end{aligned}
$$

If $Y \sim \text{beta}(p, q, a, h)$, then we can use the relation $Y = a + hX$ to obtain

$$
EY = a + hEX = a + \frac{hp}{p + q} \quad\quad (16.13)
$$

and

$$
\begin{aligned}
EY^2 &= a^2 + 2ahEX + h^2 EX^2 \\
&= a^2 + \frac{2ahp}{p + q} + \frac{h^2 p(p + 1)}{(p + q)(p + q + 1)} \;. \quad\quad (16.14)
\end{aligned}
$$

Central moments:

If $X \sim \text{beta}(p, q)$, then we readily find from (16.9) and (16.10) the variance of ξ to be

$$
\beta_2 = \text{Var } X = E(X - EX)^2 = \alpha_2 - \alpha_1^2 = \frac{pq}{(p + q)^2(p + q + 1)} \;. \quad\quad (16.15)
$$

Similarly, for the $\text{beta}(p, q, a, h)$ distributed random variable Y, we find from (16.13) and (16.14) the variance of Y to be

$$
\text{Var } Y = h^2 \, \text{Var } X = \frac{h^2 pq}{(p + q)^2(p + q + 1)} \;. \quad\quad (16.16)
$$

Proceeding in a similar fashion, when $X \sim \text{beta}(p, q)$, we can show from $(16.9) - (16.12)$ that the third and fourth central moments of X are given by

$$
\begin{aligned}
\beta_3 &= \alpha_3 - 3\alpha_2\alpha_1 + 2\alpha_1^3 \\
&= \frac{2pq(q-p)}{(p+q)^3(p+q+1)(p+q+2)}, \qquad (16.17) \\
\beta_4 &= \alpha_4 - 4\alpha_3\alpha_1 + 6\alpha_2\alpha_1^2 - 3\alpha_1^4 \\
&= \frac{3pq\left(p^2q + pq^2 + 2p^2 - 2pq + 2q^2\right)}{(p+q)^4(p+q+1)(p+q+2)(p+q+3)}, \qquad (16.18)
\end{aligned}
$$

respectively.

Exercise 16.2 Derive the expressions in (16.17) and (16.18).

Exercise 16.3 At the beginning of this chapter we considered a stick of length 1, which was broken at random into $(n+1)$ pieces, having respective lengths

$$
\begin{aligned}
D_1 &= U_{1,n}, \quad D_2 = U_{2,n} - U_{1,n}, \ldots, \quad D_k = U_{k,n} - U_{k-1,n}, \ldots, \\
D_n &= U_{n,n} - U_{n-1,n}, \quad D_{n+1} = 1 - U_{n,n}.
\end{aligned}
$$

These random variables D_k $(k = 1, 2, \ldots, n+1)$ are called *spacings* of the uniform order statistics, or simply *uniform spacings*. Then, by making use of relations (16.3) and (16.9), show that

$$
ED_k = \frac{1}{n+1}
$$

for any $1 \le k \le n+1$.

Moments of negative order.

If $X \sim \text{beta}(p, q)$, then it is easy to find moments about zero of negative order. In fact, the moment

$$
EX^{-n} = \frac{1}{B(p, q)} \int_0^1 x^{p-n-1}(1-x)^{q-1}\, dx
$$

exists if $n < p$ and is given by

$$
EX^{-n} = \frac{B(p-n, q)}{B(p, q)} = \frac{\Gamma(p-n)\Gamma(p+q)}{\Gamma(p+q-n)\Gamma(p)}. \qquad (16.19)
$$

In particular, we obtain

$$
E\left(\frac{1}{X}\right) = \frac{\Gamma(p-1)\Gamma(p+q)}{\Gamma(p+q-1)\Gamma(p)} = \frac{p+q-1}{p-1}, \qquad p > 1, \qquad (16.20)
$$

and

$$
E\left(\frac{1}{X^2}\right) = \frac{\Gamma(p-2)\Gamma(p+q)}{\Gamma(p+q-2)\Gamma(p)} = \frac{(p+q-1)(p+q-2)}{(p-1)(p-2)}, \qquad p > 2. \qquad (16.21)
$$

16.6   Shape Characteristics

Let $X \sim \text{beta}(p, q)$. Then, from the expressions in (16.15), (16.17), and (16.18), we obtain Pearson's coefficients of skewness and kurtosis of X as

$$\gamma_1 = \frac{\beta_3}{\beta_2^{3/2}} = \frac{2(q-p)}{p+q+2}\sqrt{\frac{p+q+1}{pq}}, \tag{16.22}$$

$$\gamma_2 = \frac{\beta_4}{\beta_2^2} = 3 + \frac{6(p-q)^2(p+q+1)}{pq(p+q+2)(p+q+3)} - \frac{6}{p+q+3}, \tag{16.23}$$

respectively.

From (16.22), we see immediately that the beta(p, q) distribution is negatively skewed for $q < p$ and positively skewed for $q > p$; further, the coefficient of skewness is zero when $q = p$ (in fact, the distribution is symmetric in this case).

Also, from (16.23), we see readily that the beta(p, q) distribution is platykurtic when $p = q$. More generally, the distribution is platykurtic, mesokurtic, and leptokurtic depending on whether

$$(p-q)^2(p+q+1) - pq(p+q+2)$$

$< 0, = 0$, and > 0, respectively.

16.7   Characteristic Function

If $X \sim \text{beta}(p, q)$, its characteristic function is

$$f_X(t) = Ee^{itX} = \frac{1}{B(p,q)} \int_0^1 e^{itx} x^{p-1}(1-x)^{q-1} \, dx. \tag{16.24}$$

If p and q are integers, then the RHS of (16.24) can be written as a finite sum of terms $e^{it}/(it)^k$ and $(it)^{-m}$, taken with some coefficients. For example,

$$f_X(t) = \frac{e^{it} - 1}{it}$$

if $p = 1$ and $q = 1$ [the case of the uniform $U(0, 1)$ distribution];

$$\begin{aligned}
f_X(t) &= 2\left(\frac{e^{it}}{it} - \frac{e^{it}}{(it)^2} + \frac{1}{(it)^2}\right) \\
&= 2\left(\frac{e^{it}}{it} + \frac{e^{it}}{t^2} - \frac{1}{t^2}\right)
\end{aligned}$$

if $p = 2$ and $q = 1$ [the power $Po(2, 0, 1)$ distribution], and

$$\begin{aligned}
f_X(t) &= 6\left(\frac{e^{it}}{(it)^2} - \frac{2e^{it}}{(it)^3} + \frac{1}{(it)^2} + \frac{2}{(it)^3}\right) \\
&= 6\left(2e^{it}it^3 - \frac{e^{it}}{t^2} - \frac{1}{t^2} - \frac{2}{it^3}\right)
\end{aligned}$$

if $p = 2$ and $q = 2$.

For nonintegral values of p and/or q, we can use the following expression of the characteristic function:

$$
\begin{aligned}
f_X(t) &= \frac{1}{B(p,q)} \sum_{k=0}^{\infty} \int_0^1 \frac{(it)^k x^{p+k-1}(1-x)^{q-1}}{k!} \, dx \\
&= \frac{1}{B(p,q)} \sum_{k=0}^{\infty} \frac{(it)^k B(p+k,q)}{k!} \\
&= \sum_{k=0}^{\infty} \frac{\Gamma(p+q)\Gamma(p+k)}{\Gamma(p)\Gamma(p+q+k)} \frac{(it)^k}{k!} = F(p, p+q; it), \quad (16.25)
\end{aligned}
$$

where

$$
\begin{aligned}
F(a, b; z) &= 1 + \frac{a}{b} z + \frac{1}{2!} \frac{a(a+1)}{b(b+1)} z^2 \\
&\quad + \frac{1}{3!} \frac{a(a+1)(a+2)}{b(b+1)(b+2)} z^3 + \cdots \quad (16.26)
\end{aligned}
$$

is the *Kummer confluent hypergeometric function*.

If $Y \sim \text{beta}(p, q, a, h)$, then its characteristic function is simply given by

$$
f_Y(t) = Ee^{itY} = Ee^{it(a+hX)} = e^{iat} F(p, p+q; ith). \quad (16.27)
$$

16.8   Decompositions

Since the beta distributions have bounded support, they cannot be infinitely divisible. Yet, we can give an example of a random variable X having a beta distribution which can be represented as a sum $X = X_1 + X_2 + \cdots + X_n$ of n independent terms [see relation (11.27), which is valid for $X \sim \text{beta}(1,1)$]. Indeed, in this example, the random variables $X_1, X_2, \ldots, X_n$ have different distributions. On the other hand, there are indecomposable beta distributions as well. For example, it is known that the beta(p, q) distribution is indecomposable if $p + q < 2$. Recently, Krysicki (1999) showed that if $X \sim \text{beta}(p, q)$, then for any $n = 2, 3, \ldots$, it can be represented in the following form:

$$
X \stackrel{d}{=} \left(\prod_{k=1}^{n} X_k \right)^{1/n}, \quad (16.28)
$$

where

$$
X_k \sim \text{beta}\left(\frac{p+k-1}{n}, \frac{q}{n} \right), \quad k = 1, 2, \ldots, n
$$

are independent random variables.

An analogous result was obtained by Johnson and Kotz (1990). They showed that if independent random variables $X_0, X_1, \ldots$ have a common beta(p, q) distribution, then

$$X \overset{d}{=} \sum_{j=0}^{\infty} (-1)^j \prod_{i=0}^{j} X_j$$

has a beta$(p, p + q)$ distribution.

16.9 Relationships with Other Distributions

As we mentioned earlier, beta distributions include uniform, power, and arc-sine distributions as special cases. Also, Eqs. (5.35) and (5.36) show that if V has the binomial $B(n, p)$ distribution and X has the beta$(m, n - m + 1)$ distribution, where $m \leq n$ are integers and $0 < p < 1$, then

$$P\{V \geq m\} = P\{X < p\}. \tag{16.29}$$

CHAPTER 17

ARCSINE DISTRIBUTION

17.1 Introduction

As mentioned in Chapter 16, the arcsine distribution is an important special case of the beta distribution. In this chapter we focus our attention on this special case and discuss many of its features and properties.

17.2 Notations

We say that a random variable $X \sim$ beta $\left(\frac{1}{2}, \frac{1}{2}\right)$ has the *standard arcsine distribution*. It follows from (16.4) that the pdf of the standard arcsine distribution is given by

$$p_X(x) = \frac{1}{\left\{\Gamma(\frac{1}{2})\right\}^2}\, x^{-1/2}(1-x)^{-1/2} \qquad \text{if } 0 < x < 1.$$

Since $\Gamma(\frac{1}{2}) = \sqrt{\pi}$, we obtain

$$p_X(x) = \frac{1}{\pi\sqrt{x(1-x)}}, \qquad 0 < x < 1. \tag{17.1}$$

The distribution of X is called the arcsine distribution since its cdf has the form

$$F_X(x) = \frac{2}{\pi}\arcsin\sqrt{x}, \qquad 0 \le x \le 1. \tag{17.2}$$

The density function in (17.1) is U-shaped with $p_X(x)$ tending to infinity when $x \to 0$ as well as when $x \to 1$, and its minimal value of $2/\pi$ is attained at $x = \frac{1}{2}$; see Figure 16.1.

Linear transformations $Y = a + hX$ of the random variable X give us a family of all arcsine distributions, which is just the family of beta$(\frac{1}{2}, \frac{1}{2}, a, h)$ distributions. Note that if

$$Y \sim \text{beta}\left(\frac{1}{2}, \frac{1}{2}, a, h\right), \qquad -\infty < a < \infty, \ h > 0,$$

then its cdf is given by

$$F_Y(x) = \frac{2}{\pi} \arcsin \sqrt{\frac{x-a}{h}}, \qquad a \le x \le a+h. \tag{17.3}$$

If we need to use the general arcsine density in (17.3), we denote it by $AS(a, h)$, and for the standard arcsine density in (17.1) we use the notation $X \sim AS(0, 1)$.

The importance of the arcsine distribution is due to its applications in the theory of random walks. Let us consider sums $S_0 = 0$, $S_k = Y_1 + \cdots + Y_k$, $k = 1, 2, \ldots, n$, where independent random variables $Y_1, Y_2, \ldots$ take on values -1 and 1 with equal probabilities; let us denote by V_n the number of positive sums among $S_0, S_1, \ldots, S_n$. Then, $Z_n = V_n/n$ is simply the fraction of positive sums. It turns out that as $n \to \infty$,

$$P\{Z_n \le x\} \to \frac{2}{\pi} \arcsin \sqrt{x}, \qquad 0 \le x \le 1.$$

Furthermore, let

$$T_n = \min\{k : S_k = \max(S_0, S_1, \ldots, S_n)\}$$

be the index of the maximal sum. Then, as $n \to \infty$,

$$P\left\{\frac{T_n}{n} \le x\right\} \to \frac{2}{\pi} \arcsin \sqrt{x}, \qquad 0 \le x \le 1 \tag{17.4}$$

once again.

Another similar example arises in the theory of random processes. It is well known that the standard Brownian motion $W(t)$ can be regarded as a limit process for random broken lines

$$S_n(t) = \frac{S_{[nt]}}{\sqrt{n}}, \qquad 0 < t < 1,$$

where the sums S_k are as defined above. Let δ be the time spent by $W(t)$, $0 \le t \le 1$, on the positive half-line, and τ be a random point on $[0, 1]$, where the maximum value of $W(t)$, $0 \le t \le 1$, is attained. Then

$$P\{\delta \le x\} = P\{\tau \le x\} = \frac{2}{\pi} \arcsin \sqrt{x}, \qquad 0 \le x \le 1. \tag{17.5}$$

Indeed, the arcsine distribution, being a special case of the beta distribution, inherits all the properties of that beta distribution. Note, for example, that $X \sim AS(0, 1)$ is symmetric with respect to $\frac{1}{2}$:

$$X \overset{d}{=} 1 - X. \tag{17.6}$$

17.3   Moments

Relation (17.6) enables us to simplify expressions for the moments of X.

Moments about zero:

Let $X \sim AS(0,1)$. Then, it follows from (16.8) that

$$
\begin{aligned}
\alpha_n = EX^n &= \frac{\Gamma(n + \frac{1}{2})}{\Gamma(n+1)\Gamma(\frac{1}{2})} = \frac{(n - \frac{1}{2})(n - \frac{3}{2}) \cdots \frac{1}{2}}{n!} \\
&= \frac{(2n)!}{4^n (n!)^2} = \frac{1}{4^n}\binom{2n}{n}, \qquad n = 1,2,\dots . \quad (17.7)
\end{aligned}
$$

In particular, we obtain

$$
\alpha_1 = EX = \frac{1}{2}, \tag{17.8}
$$

$$
\alpha_2 = EX^2 = \frac{3}{8}, \tag{17.9}
$$

$$
\alpha_3 = EX^3 = \frac{5}{16}, \tag{17.10}
$$

$$
\alpha_4 = EX^4 = \frac{35}{128}. \tag{17.11}
$$

Central moments:

Since $X \sim AS(0,1)$ is symmetric with respect to its mean $EX = \frac{1}{2}$, we immediately have

$$
\beta_{2n+1} = E\left(X - \frac{1}{2}\right)^{2n+1} = 0, \qquad n = 1,2,\dots .
$$

For central moments of even order, we have

$$
\begin{aligned}
\beta_{2n} &= E\left(X - \frac{1}{2}\right)^{2n} \\
&= \frac{1}{B(\frac{1}{2},\frac{1}{2})} \int_0^1 \left(x - \frac{1}{2}\right)^{2n} x^{-1/2}(1-x)^{-1/2}\, dx \\
&= \frac{2}{\{\Gamma(\frac{1}{2})\}^2} \int_{1/2}^1 \left(x - \frac{1}{2}\right)^{2n} x^{-1/2}(1-x)^{-1/2}\, dx \\
&= \frac{2}{\{\Gamma(\frac{1}{2})\}^2} \int_0^{1/2} u^{2n}\left(\frac{1}{4} - u^2\right)^{-1/2} du \\
&= \frac{1}{4^n \{\Gamma(\frac{1}{2})\}^2} \int_0^1 z^{n-1/2}(1-z)^{-1/2}\, dz \\
&= \frac{B(n + \frac{1}{2}, \frac{1}{2})}{4^n \{\Gamma(\frac{1}{2})\}^2} = \frac{\Gamma(n + \frac{1}{2})}{4^n \Gamma(n+1)\Gamma(\frac{1}{2})}. \quad (17.12)
\end{aligned}
$$

Comparing (17.7) and (17.12), we see that

$$\beta_{2n} = \frac{\alpha_n}{4^n} = \frac{(2n)!}{4^{2n}(n!)^2} = \frac{1}{4^{2n}}\binom{2n}{n}, \qquad n = 1, 2, \ldots. \qquad (17.13)$$

It readily follows from (17.13) that the variance of ξ is given by

$$\operatorname{Var} X = E(X - EX)^2 = \beta_2 = \frac{1}{8}. \qquad (17.14)$$

17.4 Shape Characteristics

As remarked earlier, the standard arcsine distribution is symmetric about its mean $\frac{1}{2}$ and, consequently, its coefficient of skewness is zero. Further, we deduce from (17.13) that the fourth central moment of X is given by

$$\beta_4 = \frac{3}{128}$$

and as a result, the Pearson coefficient of kurtosis is

$$\gamma_2 = \frac{\beta_4}{\beta_2^2} = \frac{3}{2} = 1.5.$$

Thus, we observe that the arcsine distribution is a symmetric platykurtic distribution which is U-shaped.

17.5 Characteristic Function

If $X \sim AS(0,1)$, its characteristic function is given by

$$
\begin{aligned}
f_X(t) = E e^{itX} &= \frac{1}{\pi} \int_0^1 e^{itx} x^{-1/2}(1-x)^{-1/2} \, dx \\
&= \sum_{k=0}^{\infty} \frac{\Gamma(k+\frac{1}{2})}{\Gamma(\frac{1}{2})\Gamma(k+1)} \frac{(it)^k}{k!} \\
&= \sum_{k=0}^{\infty} \frac{(k-\frac{1}{2})(k-\frac{3}{2})\cdots\frac{1}{2}}{(k!)^2} (it)^k \\
&= \sum_{k=0}^{\infty} \frac{(2k)!}{4^k(k!)^3} (it)^k \\
&= \sum_{k=0}^{\infty} \frac{1}{4^k k!} \binom{2k}{k} (it)^k. \qquad (17.15)
\end{aligned}
$$

Comparing this expression with (16.25), we see that the RHS of (17.15) can be rewritten as $F(\frac{1}{2}, 1; it)$, where $F(a, b; z)$ is the Kummer confluent hypergeometric function defined earlier in (16.26).

17.6 Relationships with Other Distributions

As mentioned already, arcsine distribution is a special case of the beta distribution. We may also recall that beta distributions with parameters $q = 1 - p$, $0 < p < 1$, are called *generalized arcsine distributions*. Indeed, the arcsine distribution is a special case ($p = q = \frac{1}{2}$) of generalized arcsine distributions. Using the probability integral transformation, we have the random variables $X \sim AS(0,1)$ and $U \sim U(0,1)$ satisfying the relation

$$\arcsin \sqrt{X} \stackrel{d}{=} \frac{\pi U}{2} . \tag{17.16}$$

Exercise 17.1 Show that if $Y \sim U(0, \pi)$, then

$$\cos Y \sim AS(-1, 2). \tag{17.17}$$

17.7 Characterizations

If $Y \sim U(0, \pi)$, it is easy to verify that $\cos^2 U$ and $(1 + \cos U)/2$ have the same distribution, meaning that

$$V^2 \stackrel{d}{=} \frac{1+V}{2} , \tag{17.18}$$

when $V \sim AS(-1, 2)$. Note that (17.18) is equivalent to the relation

$$4\left(X - \frac{1}{2}\right)^2 \stackrel{d}{=} X \tag{17.19}$$

when $X \sim AS(0, 1)$.

Exercise 17.2 Show that (17.19) is valid for the standard arcsine distribution $AS(0, 1)$, comparing central moments β_{2n} and moments about zero α_n ($n = 1, 2, \ldots$) given in (17.13) and (17.7), respectively.

Note that property (17.19) characterizes $AS(0, 1)$ distribution while the relation (17.18) characterizes $AS(-1, 2)$ distribution.

17.8   Decompositions

As mentioned in Chapter 16, beta(p, q) distributions with $p + q < 2$ are indecomposable. Hence, the arcsine distribution is indecomposable; that is, we cannot find two nondegenerate independent random variables U and V such that $U + V \sim AS(0, 1)$. However, the following relation holds for any $n = 2, 3, \ldots$ [see Krysicki (1999)]:

$$X \overset{d}{=} \left(\prod_{k=1}^{n} X_k \right)^{1/n}, \tag{17.20}$$

where $X \sim AS(0, 1)$ and $X_k \sim \text{beta}\left(\dfrac{2k - 1}{2n}, \dfrac{1}{2n} \right)$, $k = 1, 2, \ldots, n$, are independent random variables. Hence, X cannot be presented as a sum of independent variables while it can be represented as a product of independent beta random variables.

CHAPTER 18

EXPONENTIAL DISTRIBUTION

18.1 Introduction

When we discussed distributions of the uniform order statistics $U_{k,n}$ $(1 \le k \le n)$ in Chapter 11, it was shown that (as $n \to \infty$)

$$\begin{aligned} F_n(x) &= P\{nU_{1,n} \le x\} \\ &= 1 - \left(1 - \frac{x}{n}\right)^n \to 1 - e^{-x}, \qquad x \ge 0, \end{aligned}$$

which is the standard exponential distribution function.

A book-length account of exponential distributions, discussing in great detail their various properties and applications, is available [Balakrishnan and Basu (1995)].

18.2 Notations

We say that a random variable X has the standard exponential distribution if its cdf $F_X(x)$ is given by

$$F_X(x) = \begin{cases} 1 - e^{-x} & \text{if } x \ge 0 \\ 0 & \text{if } x < 0. \end{cases} \tag{18.1}$$

The corresponding probability density function has the form

$$p_X(x) = \begin{cases} e^{-x} & \text{if } x \ge 0 \\ 0 & \text{if } x < 0. \end{cases} \tag{18.2}$$

Linear transformations $Y = a + \lambda X$, $-\infty < a < \infty$, $\lambda > 0$, enlarge the family of exponential distributions to those with cdf

$$F_Y(x) = \begin{cases} 1 - \exp\left(-\dfrac{x - a}{\lambda}\right) & \text{if } x \ge a \\ 0 & \text{if } x < a. \end{cases} \tag{18.3}$$

Note that $F_Y(x)$ in (18.3) can be rewritten as

$$F_Y(x) = \max\left\{0, 1 - \exp\left(-\frac{x-a}{\lambda}\right)\right\}, \quad -\infty < x < \infty. \tag{18.4}$$

The corresponding pdf is given by

$$p_Y(x) = \begin{cases} \frac{1}{\lambda} \exp\left\{-\frac{x-a}{\lambda}\right\} & \text{if } x \geq a \\ 0 & \text{if } x < a. \end{cases} \tag{18.5}$$

In the sequel, we will use the notation $Y \sim E(a, \lambda)$ to denote a random variable Y having the exponential distribution with location parameter a and scale parameter $\lambda > 0$. In many situations, we deal with nonshifted ($a = 0$) exponential distribution $E(0, \lambda)$. For the sake of simplicity, we will denote it by $Y \sim E(\lambda)$, and in this case

$$F_Y(x) = \max\left\{0, 1 - \exp\left(-\frac{x}{\lambda}\right)\right\} \tag{18.6}$$

and

$$p_Y(x) = \frac{1}{\lambda} \exp\left(-\frac{x}{\lambda}\right) \quad \text{if } x \geq 0. \tag{18.7}$$

Note that if $X \sim E(1)$, then $Y = \lambda X \sim E(\lambda)$.

Exercise 18.1 Let $X \sim E(1)$ and $Y \sim E(\lambda)$ be independent random variables. Then, find the value of the parameter λ such that $P\{X \geq Y\} = \frac{1}{3}$.

Exercise 18.2 Let X and Y be independent standard exponential random variables. Find the distribution of X/Y.

18.3   Laplace Transform and Characteristic Function

If $Y \sim E(\lambda)$, then its Laplace transform $\varphi_Y(s) = Ee^{-sY}$ has the following form:

$$\varphi_Y(s) = \frac{1}{\lambda}\int_0^\infty \exp\left(-\frac{\lambda sx + x}{\lambda}\right) dx = \frac{1}{1 + \lambda s}. \tag{18.8}$$

If $V = a + Y$, then $V \sim E(a, \lambda)$, and consequently,

$$\varphi_V(s) = Ee^{-sV} = e^{-as}Ee^{-sV} = \frac{e^{-as}}{1 + \lambda s}. \tag{18.9}$$

To obtain the characteristic function of an exponentially distributed random variable, we can use the following relation between Laplace transforms $\varphi_X(s) = Ee^{-sX}$ and characteristic functions $f_X(t) = Ee^{itX}$:

$$f_X(t) = \varphi_X(-it). \qquad (18.10)$$

Using (18.10) and the expression for the Laplace transform of exponential distribution given above, we can write the characteristic functions of $X \sim E(1)$, $Y \sim E(\lambda)$, and $V \sim E(a, \lambda)$ as

$$f_X(t) = \frac{1}{1 - it}, \qquad (18.11)$$

$$f_Y(t) = \frac{1}{1 - i\lambda t}, \qquad (18.12)$$

$$f_V(t) = \frac{e^{iat}}{1 - i\lambda t}, \qquad (18.13)$$

respectively.

18.4 Moments

The exponential decay of the pdf (18.5) provides the existence of all the moments of exponential distribution.

Moments about zero:
If $Y \sim E(\lambda)$, then

$$\begin{aligned}
\alpha_n = EY^n &= \frac{1}{\lambda} \int_0^\infty x^n \exp\left(-\frac{x}{\lambda}\right) dx \\
&= \lambda^n \int_0^\infty x^n e^{-x}\, dx = \lambda^n \Gamma(n+1) \\
&= \lambda^n n!, \qquad n = 1, 2, \ldots. \qquad (18.14)
\end{aligned}$$

In particular, we have

$$\begin{aligned}
EY &= \lambda, & (18.15) \\
EY^2 &= 2\lambda^2, & (18.16) \\
EY^3 &= 6\lambda^3, & (18.17) \\
EY^4 &= 24\lambda^4. & (18.18)
\end{aligned}$$

In the general case when $V \sim E(a, \lambda)$, we obtain

$$\begin{aligned}
EV^n = E(a + Y)^n &= \sum_{r=0}^n \binom{n}{r} a^{n-r} EY^r \\
&= n! \sum_{r=0}^n \frac{\lambda^r a^{n-r}}{(n-r)!}, \qquad n = 1, 2, \ldots. \quad (18.19)
\end{aligned}$$

Central moments:

Indeed, the central moments coincide for random variables $V \sim E(a, \lambda)$ and $Y \sim E(\lambda)$. Let X have the standard exponential $E(1)$ distribution. Then

$$V - EV \stackrel{d}{=} Y - EY \stackrel{d}{=} \lambda(X - EX),$$

and therefore

$$
\begin{aligned}
\beta_n = E(V - EV)^n &= E(Y - \lambda)^n = \lambda^n E(X - 1)^n \\
&= \lambda^n \sum_{r=0}^{n} \binom{n}{r} (-1)^{n-r} EX^r \\
&= \lambda^n n! \sum_{r=0}^{n} \frac{(-1)^{n-r}}{(n-r)!} \\
&= \lambda^n n! \sum_{r=0}^{n} \frac{(-1)^r}{r!}, \qquad n = 1, 2, \dots. \quad (18.20)
\end{aligned}
$$

We can show from (18.20) that central moments β_n satisfy the following recurrence relations:

$$
\begin{aligned}
\beta_1 &= 0, \\
\beta_{n+1} &= (n+1)\lambda \beta_n + (-1)^{n+1} \lambda^{n+1}, \qquad n = 1, 2, \dots. \quad (18.21)
\end{aligned}
$$

In particular, we obtain from (18.20) the variance of Y to be

$$\text{Var } Y = \beta_2 = \lambda^2. \tag{18.22}$$

18.5   Shape Characteristics

Let $Y \sim E(\lambda)$. Then, from (18.7), we see readily that the distribution is unimodal with the mode at 0. It has an exponential decay form. Also, from (18.20), we obtain the third and fourth central moments of Y as

$$\beta_3 = 2\lambda^3 \quad \text{and} \quad \beta_4 = 9\lambda^4.$$

We then readily find Pearson's measures of skewness and kurtosis as

$$\gamma_1 = 2 \quad \text{and} \quad \gamma_2 = 9,$$

respectively. Thus, the exponential distribution is a positively skewed and leptokurtic distribution which is reverse *J*-shaped.

Plots of exponential density function are presented in Figure 18.1 for some choices of λ.

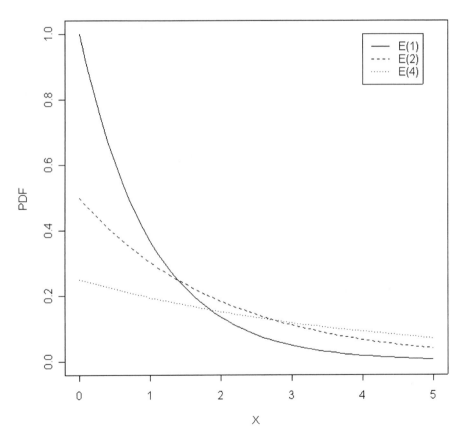

Figure 18.1. Plots of exponential density function

18.6 Entropy

Since the random variable $V \sim E(a, \lambda)$ has pdf as given in (18.5), its entropy $H(V)$ is defined by

$$H(V) = \int_a^\infty \frac{1}{\lambda} \exp\left\{-\frac{x-a}{\lambda}\right\} \left(\log \lambda + \frac{x-a}{\lambda}\right) dx = 1 + \log \lambda. \quad (18.23)$$

Consider the set of all probability density functions $p(x)$ satisfying the following restrictions:

(a) $p(x) > 0$ for $x \geq 0$ and $p(x) = 0$ for $x < 0$;
(b) $\int_0^\infty xp(x)\, dx = C$, where C is a positive constant.

It happens that the maximal value of entropy for this set of pdf's is attained for

$$p(x) = \frac{1}{C} \exp\left(-\frac{x}{C}\right), \qquad x \geq 0,$$

that is, for an exponential distribution with mean C.

18.7 Distributions of Minima

Let $Y_k \sim E(\lambda_k)$, $k = 1, 2, \ldots, n$, be independent random variables, and $m_n = \min\{Y_1, \ldots, Y_n\}$.

Exercise 18.3 Prove that m_n (for any $n = 1, 2, \ldots$) also has the exponential $E(\lambda)$ distribution, where

$$\lambda = \lambda_1 + \cdots + \lambda_n.$$

The statement of Exercise 18.1 enables us to show that

$$\min\{Y_1, \ldots, Y_a\} \overset{d}{=} \frac{Y_1}{n}, \qquad n = 1, 2, \ldots \quad (18.24)$$

when $Y_1, Y_2, \ldots$ are independent and identically distributed as $E(\lambda)$. It is of interest to mention that property (18.24) characterizes the exponential $E(\lambda)$ ($\lambda > 0$) distribution.

18.8   Uniform and Exponential Order Statistics

In Chapter 11 [see (11.39)] we introduced the uniform order statistics

$$U_{1,n} \leq U_{2,n} \leq \cdots \leq U_{n,n}$$

arising from independent and identically distributed random variables $U_k \sim U(0,1)$, $k = 1, 2, \ldots, n$. The analogous ordering of independent exponential $E(1)$ random variables $X_1, X_2, \ldots, X_n$ gives us the *exponential order statistics*

$$X_{1,n} \leq X_{2,n} \leq \cdots \leq X_{n,n}.$$

Note that

$$X_{1,n} = \min\{X_1, X_2, \ldots, X_n\}$$

and

$$X_{n,n} = \max\{X_1, X_2, \ldots, X_n\} \ .$$

There exist useful representations of uniform as well as exponential order statistics in terms of sums of independent exponential random variables.

To be specific, let $X_1, X_2, \ldots$ be independent exponential $E(1)$ random variables, and

$$S_k = X_1 + X_2 + \cdots + X_k, \qquad k = 1, 2, \ldots.$$

Then, the following relations are valid for any $n = 1, 2, \ldots$:

$$\{U_{k,n}\}_{k=1}^n \stackrel{d}{=} \left\{ \frac{S_k}{S_{n+1}} \right\}_{k=1}^n \tag{18.25}$$

and

$$\{X_{k,n}\}_{k=1}^n \stackrel{d}{=} \left\{ \frac{X_1}{n} + \frac{X_2}{n-1} + \cdots + \frac{X_k}{n-k+1} \right\}_{k=1}^n . \tag{18.26}$$

Note that relation (18.24) is a special case of (18.26). The distributional relations in (18.25) and (18.26) enable us to obtain some interesting corollaries. For example, it is easy to see that the *exponential spacings*

$$D_{1,n} = X_{1,n}, \quad D_{2,n} = X_{2,n} - X_{1,n}, \ldots, \quad D_{n,n} = X_{n,n} - X_{n-1,n}$$

are independent, and that

$$(n - k + 1)D_{k,n} \sim E(1).$$

It also follows from (18.26) that

$$
\begin{aligned}
EX_{k,n} &= E\left(\frac{X_1}{n} + \frac{X_2}{n-1} + \cdots + \frac{X_k}{n-k+1} \right) \\
&= \frac{1}{n} + \frac{1}{n-1} + \cdots + \frac{1}{n-k+1}, \quad 1 \leq k \leq n. \tag{18.27}
\end{aligned}
$$

Upon setting $k = n = 1$ in (18.25), we obtain the following interesting result: If X_1 and X_2 are independent exponential $E(1)$ random variables, then $Z = X_1/(X_1 + X_2)$ has the uniform $U(0,1)$ distribution.

Exercise 18.4 Using the representation in (18.26), show that the variance of $X_{k,n}$ is given by

$$\text{Var } X_{k,n} = \frac{1}{n^2} + \frac{1}{(n-1)^2} + \cdots + \frac{1}{(n-k+1)^2}, \qquad 1 \le k \le n,$$

and that the covariance of $X_{k,n}$ and $X_{\ell,n}$ is simply

$$\text{Cov}(X_{k,n}, X_{\ell,n}) = \text{Var } X_{k,n}, \qquad 1 \le k < \ell \le n .$$

18.9 Convolutions

Let $Y_1 \sim E(\lambda_1)$ and $Y_2 \sim E(\lambda_2)$ be independent random variables, and $V = Y_1 + Y_2$.

Exercise 18.5 Show that the pdf $p_V(x)$ of V has the following form:

$$p_V(x) = \frac{e^{-x/\lambda_1} - e^{-x/\lambda_2}}{\lambda_1 - \lambda_2}, \qquad x \ge 0 \tag{18.28}$$

if $\lambda_1 \ne \lambda_2$, and

$$p_V(x) = \frac{x}{\lambda^2} e^{-x/\lambda}, \qquad x \ge 0 \tag{18.29}$$

if $\lambda_1 = \lambda_2 = \lambda$.

Equation (18.29) gives the distribution of the sum of two independent random variables having the same exponential distribution. Now, let us consider the sum

$$V_n = X_1 + \cdots + X_n$$

of n independent exponential random variables. For the sake of simplicity, we will suppose that $X_k \sim E(1)$, $k = 1, 2, \ldots, n$. It follows from (18.8) that the Laplace transform $\varphi(s)$ of any X_k has the form

$$\varphi(s) = \int_0^\infty e^{-sx} e^{-x} \, dx = \frac{1}{1+s} . \tag{18.30}$$

Then the Laplace transform $\varphi_n(s)$ of the sum V_n, which also takes on only positive values, is given by

$$\varphi_n(s) = \int_0^\infty e^{-sx} p_n(x) \, dx = (1+s)^{-n}, \tag{18.31}$$

where $p_n(x)$ denotes the pdf of V_n. Comparing (18.30) and (18.31), we can readily see that

$$\varphi_n(s) = \frac{(-1)^{n-1}}{(n-1)!} \, \varphi^{(n-1)}(s)$$

and, hence,

$$
\begin{aligned}
\int_0^\infty e^{-sx} p_n(x) \, dx &= \frac{(-1)^{n-1}}{(n-1)!} \int_0^\infty e^{-sx}(-x)^{n-1} e^{-x} \, dx \\
&= \frac{1}{(n-1)!} \int_0^\infty e^{-sx} x^{n-1} e^{-x} \, dx.
\end{aligned}
\tag{18.32}
$$

It then follows from (18.32) that

$$p_n(x) = \frac{1}{(n-1)!} \, x^{n-1} e^{-x}, \qquad x \geq 0. \tag{18.33}$$

Probability density function in (18.33) is a special case of *gamma distributions*, which is discussed in detail in Chapter 20.

Exercise 18.6 Let X_1 and X_2 be independent random variables having the standard $E(1)$ distribution, and $W = X_1 - X_2$. Show that the pdf of W has the form

$$p_W(x) = \frac{1}{2} \, e^{-|x|}, \qquad -\infty < x < \infty. \tag{18.34}$$

The probability density function in (18.34) corresponds to the *standard Laplace distribution*, which is discussed in detail in Chapter 19.

18.10   Decompositions

Let $X \sim E(1)$. We will now present two different ways to express X as a sum of two independent nondegenerate random variables.

(1) Consider random variables $V = [X]$ and $U = \{X\}$, which are the integer and fractional parts of X, respectively; for example, $V = n$ and $U = x$ if $X = n + x$, where $n = 0, 1, 2, \ldots$ and $0 \leq x < 1$. It is evident that

$$X = V + U. \tag{18.35}$$

Let us show that the terms on the RHS of (18.35) are independent. It is not difficult to check that V takes on values $0, 1, 2, \ldots$ with probabilities

$$p_n = P\{V = n\} = P\{n \leq X < n+1\} = (1 - \lambda)\lambda^n, \tag{18.36}$$

where $\lambda = 1/e$. This simply means that V has the geometric $G(1/e)$ distribution. We see that the fractional part of X takes values on the interval $[0, 1)$ and

$$
\begin{aligned}
F_U(x) &= P\{U \le x\} = \sum_{n=0}^{\infty} P\{n \le X \le n + x\} \\
&= \sum_{n=0}^{\infty} e^{-n}(1 - e^{-x}) = \frac{1 - e^{-x}}{1 - e^{-1}}, \qquad 0 \le x \le 1. (18.37)
\end{aligned}
$$

To check the independence of random variables V and U, we must show that for any $n = 0, 1, 2, \ldots$ and $0 \le x \le 1$, the following condition holds:

$$
P\{V = n, \ U \le x\} = p_n F_U(x). \tag{18.38}
$$

Relation (18.38) becomes evident once we note that

$$
P\{V = n, \ U \le x\} = P\{n \le X < n + x\} = e^{-n}(1 - e^{-x}).
$$

Thus, we have expressed X having the standard exponential distribution as a sum of two independent random variables. Representation (18.35) is also valid for $Y \sim E(\lambda)$, $\lambda > 0$. In the general case, when $Y \sim E(a, \lambda)$, the following general form of (18.35) holds:

$$
Y = a + [Y - a] + \{Y - a\},
$$

where the integer and fractional parts ($[Y - a]$ and $\{Y - a\}$) of the random variable $Y - a$ are independent.

(2) Along with the random variable $X \sim E(1)$, let us consider two independent random variables Y_1 and Y_2 with a common pdf:

$$
g(x) = \frac{1}{\sqrt{\pi}} \, x^{-1/2} \, e^{-x}, \qquad x > 0. \tag{18.39}
$$

It is easy to prove that the nonnegative function $g(x)$ above is really a probability density function because of the fact that

$$
\int_0^{\infty} g(x) \, dx = \frac{1}{\sqrt{\pi}} \int_0^{\infty} x^{-1/2} e^{-x} \, dx = \frac{\Gamma(\frac{1}{2})}{\sqrt{\pi}} = 1. \tag{18.40}
$$

Exercise 18.7 Show that the sum $Y_1 + Y_2$ has the exponential $E(1)$ distribution.

Thus, we have established that the exponential distributions are decomposable. Moreover, in Chapter 20 we show that any exponential distribution is infinitely divisible.

18.11   Lack of Memory Property

In Chapter 6 [see (6.39)], we established that the geometric $G(p)$ random variable X for any $n = 0, 1, \ldots$ and $m = 0, 1, \ldots$ satisfies the following equality:

$$P\{X \geq n + m \mid X \geq m\} = P\{X \geq n\}.$$

Furthermore, among all discrete distributions taking on integer values, geometric distribution is the only distribution that possesses this "lack of memory" property.

Now, let us consider $Y \sim E(\lambda)$, $\lambda > 0$. It is not difficult to see that the equality

$$P\{Y \geq x + y \mid Y \geq y\} = P\{Y \geq x\} \tag{18.41}$$

holds for any $x \geq 0$ and $y \geq 0$. This "lack of memory" property characterizes the exponential $E(\lambda)$ distribution; that is, if a nonnegative random variable Y with cdf $F(x)$ [such that $F(x) < 1$ for any $x > 0$] satisfies (18.41), then

$$F(x) = 1 - e^{-x/\lambda}, \qquad x > 0,$$

for some positive constant λ.

CHAPTER 19

LAPLACE DISTRIBUTION

19.1 Introduction

In Chapter 18 (see Exercise 18.4) we considered the difference $V = X_1 - X_2$ of two independent random variables with X_1 and X_2 having the standard exponential $E(1)$ distribution. It was mentioned there that the pdf of V is of the form

$$p_V(x) = \frac{1}{2} e^{-|x|}, \qquad -\infty < x < \infty.$$

This distribution is one of the earliest in probability theory, and it was introduced by Laplace (1774).

A book-length account of Laplace distributions, discussing in great detail their various properties and applications, is available and is due to Kotz et al. (2001).

19.2 Notations

We say that a random variable X has the *Laplace distribution* if its pdf $p_X(x)$ is given by

$$p_X(x) = \frac{1}{2\lambda} \exp\left(-\frac{|x-a|}{\lambda}\right), \qquad -\infty < x < \infty. \qquad (19.1)$$

We use $X \sim L(a, \lambda)$ to indicate that X has the Laplace distribution in (19.1) with a location parameter a and a scale parameter λ $(-\infty < a < \infty, \lambda > 0)$. In the special case when X has a symmetric distribution with the pdf

$$p_X(x) = \frac{1}{2\lambda} \exp\left(-\frac{|x|}{\lambda}\right), \qquad -\infty < x < \infty, \qquad (19.2)$$

we denote it by $X \sim L(\lambda)$ for the sake of simplicity. For instance, $V \sim L(1)$ denotes that V has the standard Laplace distribution with pdf

$$p_V(x) = \frac{1}{2} e^{-|x|}, \qquad -\infty < x < \infty, \qquad (19.3)$$

and its cdf $F_V(x)$ has the form

$$F_V(x) = \begin{cases} \frac{1}{2}\, e^x & \text{if } x < 0 \\ 1 - \frac{1}{2}\, e^{-x} & \text{if } x \geq 0. \end{cases} \tag{19.4}$$

Indeed, if $V \sim L(1)$, then $Y = \lambda V \sim L(\lambda)$, and $X = a + \lambda V \sim L(a, \lambda)$.

Laplace distributions are also known under different names in the literature: the *first law of Laplace* (the second law of Laplace is the standard normal distribution), *double exponential* (the name *double exponential* or, sometimes, *doubly exponential* is also applied to the distribution having the cdf

$$F(x) = \exp(-e^{-x}), \qquad -\infty < x < \infty,$$

which is one of the three types of extreme value distributions), *two-tailed exponential,* and *bilateral exponential* distributions.

19.3   Characteristic Function

Recall that if $V \sim L(1)$, it can be represented as $V = X_1 - X_2$, where X_1 and X_2 are independent random variables and have standard exponential $E(1)$ distribution. Consequently, the characteristic function of V is

$$f_V(t) = Ee^{itV} = Ee^{itX_1} Ee^{-itX_2}$$

and, therefore, we obtain from (18.11) that

$$f_V(t) = \frac{1}{(1 - it)(1 + it)} = \frac{1}{1 + t^2} \,. \tag{19.5}$$

Since the linear transformation $X = a + \lambda V$ leads to the Laplace $L(a, \lambda)$ distribution, we readily get

$$f_X(t) = \frac{e^{iat}}{1 + \lambda^2 t^2} \,. \tag{19.6}$$

It is of interest to mention here that (19.5) gives a simple way to obtain the characteristic function of the Cauchy distribution [see also (12.6) and (12.7)]. In fact, (19.5) is equivalent to the following inverse Fourier transform, which relates the characteristic function $f_V(t)$ with the probability density function $p_V(x)$:

$$\frac{1}{2}\, e^{-|x|} = \frac{1}{2\pi} \int_{-\infty}^{\infty} \frac{e^{-itx}}{1 + t^2}\, dt \,. \tag{19.7}$$

It then follows from (19.7) that

$$\int_{-\infty}^{\infty} e^{itx}\, \frac{1}{\pi(1 + x^2)}\, dx = e^{-|t|}. \tag{19.8}$$

Now we can see from (19.8) that the Cauchy $C(0, 1)$ distribution with pdf

$$p(x) = \frac{1}{\pi(1 + x^2)}$$

has characteristic function

$$f(t) = e^{-|t|}.$$

19.4   Moments

The exponential rate of the decreasing nature of the pdf in (19.1) entails the existence of all the moments of the Laplace distribution.

Moments about zero:
   Let $Y \sim L(\lambda)$. Since Y has a symmetric distribution, we readily have

$$\alpha_{2n-1} = EY^{2n-1} = 0, \qquad n = 1, 2, \ldots \, .$$

For moments of even order, we have

$$\begin{aligned} \alpha_{2n} = EY^{2n} &= \frac{1}{2\lambda} \int_{-\infty}^{\infty} x^{2n} \exp\left(-\frac{|x|}{\lambda}\right) dx \\ &= \frac{1}{\lambda} \int_{0}^{\infty} x^{2n} \exp\left(-\frac{x}{\lambda}\right) dx \\ &= \lambda^{2n} \int_{0}^{\infty} x^{2n} e^{-x} \, dx \\ &= \lambda^{2n} \, \Gamma(2n+1) = \lambda^{2n} \, (2n)!, \qquad n = 1, 2, \ldots . \end{aligned} \quad (19.9)$$

In particular, we have

$$EY^2 = 2\lambda^2 \qquad\qquad (19.10)$$

and

$$EY^4 = 24\lambda^4. \qquad\qquad (19.11)$$

   In the general case when $X = a + Y \sim L(a, \lambda)$, we also have

$$EX^n = E(a+Y)^n = \sum_{r=0}^{n} \binom{n}{r} a^{n-r} EY^r, \quad n = 1, 2, \ldots, \qquad (19.12)$$

where

$$EY^r = \begin{cases} 0 & \text{if } r = 1, 3, 5, \ldots \\ \lambda^r r! & \text{if } r = 0, 2, 4, \ldots . \end{cases}$$

Central moments:
   Let $Y \sim L(\lambda)$ and $X = a + Y \sim L(a, \lambda)$. It is clear that central moments of X and moments about zero of Y are the same since

$$\beta_n = E(X - EX)^n = E(Y - EY)^n = EY^n, \qquad n = 1, 2, \ldots . \qquad (19.13)$$

From (19.13) and (19.10), we immediately find that

$$\mathrm{Var}\, Y = \beta_2 = 2\lambda^2. \qquad\qquad (19.14)$$

19.5   Shape Characteristics

Let $Y \sim L(\lambda)$. Then, due to the symmetry of the distribution of Y, we readily have the Pearson's coefficient of skewness as

$$\gamma_1 = 0.$$

Further, from (19.11), we find the fourth central moment of Y as

$$
\begin{aligned}
\beta_4 &= \alpha_4 - 4\alpha_3\alpha_1 + 6\alpha_2\alpha_1^2 - 3\alpha_1^4 \\
&= \alpha_4 = 24\lambda^4.
\end{aligned}
\tag{19.15}
$$

From Eqs. (19.15) and (19.14), we readily find the Pearson's coefficient of kurtosis as

$$\gamma_2 = \frac{\beta_4}{\beta_2^2} = 6.
\tag{19.16}$$

Thus, we find the Laplace distribution to be a symmetric leptokurtic distribution.

19.6   Entropy

The entropy of the random variable $X \sim L(a, \lambda)$ is given by

$$
\begin{aligned}
H(X) &= \int_{-\infty}^{a} \frac{1}{2\lambda} \exp\left\{\frac{x-a}{\lambda}\right\} \left(\log(2\lambda) - \frac{x-a}{\lambda}\log e\right) dx \\
&\quad + \int_{a}^{\infty} \frac{1}{2\lambda} \exp\left\{-\frac{x-a}{\lambda}\right\} \left(\log(2\lambda) + \frac{x-a}{\lambda}\log e\right) dx \\
&= \int_{-\infty}^{0} \frac{1}{2} e^{x} \left\{\log(2\lambda) - x\log e\right\} dx \\
&\quad + \int_{0}^{\infty} \frac{1}{2} e^{-x} \left\{\log(2\lambda) + x\log e\right\} dx \\
&= \log(2\lambda e) = 1 + \log(2\lambda).
\end{aligned}
\tag{19.17}
$$

Consider the set of all probability density functions $p(x)$ satisfying the following restriction:

$$\int_{-\infty}^{\infty} |x|p(x)\, dx = C,
\tag{19.18}$$

where C is a positive constant. Then, it has been proved that the maximal value of entropy for this set of densities is attained if

$$p(x) = \frac{1}{2C} \exp\left(-\frac{|x|}{C}\right),$$

that is, for a Laplace distribution.

19.7 Convolutions

Let $Y_1 \sim L(\lambda)$ and $Y_2 \sim L(\lambda)$ be independent random variables, $W = Y_1 + Y_2$, and $T = Y_1 - Y_2$. Since the Laplace $L(\lambda)$ distribution is symmetric about zero for any $\lambda > 0$, it is clear that the random variables W and T have the same distribution.

Exercise 19.1 Show that densities $p_W(x)$ and $p_T(x)$ of the random variables W and T have the following form:

$$p_W(x) = p_T(x) = \frac{1}{4\lambda} \left(1 + \frac{|x|}{\lambda} \right) \exp \left(-\frac{|x|}{\lambda} \right), \quad -\infty < x < \infty. \quad (19.19)$$

Next, let $Y_1 \sim L(\lambda_1)$ and $Y_2 \sim L(\lambda_2)$ be independent random variables each having Laplace distribution with different scale parameters. In this case, too, the random variables $W = Y_1 + Y_2$ and $T = Y_1 - Y_2$ have a common distribution. To find this distribution, we must recall that characteristic functions $f_1(t) = Ee^{itY_1}$ and $f_2(t) = Ee^{itY_2}$ have the form

$$f_k(t) = \frac{1}{1 + \lambda_k^2 t^2}, \qquad k = 1, 2,$$

and hence the characteristic function $f_W(t)$ is given by

$$
\begin{aligned}
f_W(t) = Ee^{itW} &= Ee^{it(Y_1 + Y_2)} \\
&= f_1(t) f_2(t) \\
&= \frac{1}{(1 + \lambda_1^2 t^2)(1 + \lambda_2^2 t^2)} \\
&= \frac{\lambda_1^2}{\lambda_1^2 - \lambda_2^2} \frac{1}{1 + \lambda_1^2 t^2} - \frac{\lambda_2^2}{\lambda_1^2 - \lambda_2^2} \frac{1}{1 + \lambda_2^2 t^2} \\
&= \frac{\lambda_1^2}{\lambda_1^2 - \lambda_2^2} f_1(t) - \frac{\lambda_2^2}{\lambda_1^2 - \lambda_2^2} f_2(t). \quad (19.20)
\end{aligned}
$$

It follows from (19.20) that the probability density functions

$$p_W(x), \quad p_1(x) = \frac{1}{2\lambda_1} \exp \left(-\frac{|x|}{\lambda_1} \right), \quad p_2(x) = \frac{1}{2\lambda_2} \exp \left(-\frac{|x|}{\lambda_2} \right)$$

of the random variables W, Y_1, and Y_2 satisfy the relationship

$$p_W(x) = \frac{\lambda_1^2}{\lambda_1^2 - \lambda_2^2} p_1(x) - \frac{\lambda_2^2}{\lambda_1^2 - \lambda_2^2} p_2(x). \quad (19.21)$$

Hence, we obtain

$$p_W(x) = \frac{1}{2} \left\{ \frac{\lambda_1}{\lambda_1^2 - \lambda_2^2} \exp \left(-\frac{|x|}{\lambda_1} \right) - \frac{\lambda_2}{\lambda_1^2 - \lambda_2^2} \exp \left(-\frac{|x|}{\lambda_2} \right) \right\},$$
$$-\infty < x < \infty. \quad (19.22)$$

19.8   Decompositions

Coming back to the representation $V = X_1 - X_2$, where $V \sim L(1)$ and independent random variables X_1 and X_2 have the standard exponential $E(1)$ distribution, we easily obtain that the distribution of V is decomposable. Recall from Chapter 18 that the random variables X_1 and X_2 are both decomposable [see, for example, (18.35) and Exercise 18.5]. This simply means that there exist independent random variables $Y_1, Y_2, Y_3,$ and Y_4 such that

$$X_1 \stackrel{d}{=} Y_1 + Y_2$$

and

$$X_2 \stackrel{d}{=} Y_3 + Y_4,$$

and hence

$$V \stackrel{d}{=} (Y_1 - Y_3) + (Y_2 - Y_4), \tag{19.23}$$

where the differences $Y_1 - Y_3$ and $Y_2 - Y_4$ are independent random variables and have nondegenerate distributions. Therefore, $V \sim L(1)$ is a decomposable random variable. Furthermore, a linear transformation $a + \lambda V$ preserves this property and so the Laplace $L(a, \lambda)$ distribution is also decomposable.

In addition, since exponential distributions are infinitely divisible, the representation $V = X_1 - X_2$ enables us to note that Laplace distributions are also infinitely divisible.

19.9   Order Statistics

Let $V_1, V_2, \ldots, V_n$ be independent and identically distributed as standard Laplace $L(1)$ distribution. Further, let $V_{1,n} < V_{2,n} < \cdots < V_{n,n}$ be the corresponding order statistics. Then, using the expressions of $p_V(x)$ and $F_V(x)$ in (19.3) and (19.4), respectively, we can write the kth moment of $V_{r,n}$ (for $1 \le r \le n$) as

$$E\left(V_{r,n}^k\right)$$

$$= \frac{n!}{(r-1)!\,(n-r)!} \int_{-\infty}^{\infty} x^k \left\{F_V(x)\right\}^{r-1} \left\{1 - F_V(x)\right\}^{n-r} p_V(x)\, dx$$

$$= \frac{n!}{(r-1)!\,(n-r)!} \left[\int_{-\infty}^{0} x^k \left\{\frac{1}{2}e^x\right\}^{r-1} \left\{1 - \frac{1}{2}e^x\right\}^{n-r} \frac{1}{2}e^x\, dx \right.$$

$$\left. + \int_{0}^{\infty} x^k \left\{1 - \frac{1}{2}e^{-x}\right\}^{r-1} \left\{\frac{1}{2}e^{-x}\right\}^{n-r} \frac{1}{2}e^{-x}\, dx \right]$$

$$= \frac{n!}{2^n(r-1)!\,(n-r)!} \left[\int_{0}^{\infty} x^k \left\{2 - e^{-x}\right\}^{r-1} \left\{e^{-x}\right\}^{n-r} e^{-x}\, dx \right.$$

$$\left. + (-1)^k \int_{0}^{\infty} x^k \left\{e^{-x}\right\}^{r-1} \left\{2 - e^{-x}\right\}^{n-r} e^{-x}\, dx \right]. \tag{19.24}$$

Now, upon writing $2-e^{-x}$ in the two integrands as $1+(1-e^{-x})$ and expanding the corresponding terms binomially, we readily obtain

$$
\begin{aligned}
& E\left(V_{r,n}^k\right) \\
& = (-1)^k \frac{n!}{2^n(r-1)!\,(n-r)!} \sum_{j=0}^{n-r}\binom{n-r}{j}\int_0^\infty x^k \left\{1-e^{-x}\right\}^j \\
& \qquad \times \left\{e^{-x}\right\}^{r-1} e^{-x}\,dx \\
& + \frac{n!}{2^n(r-1)!\,(n-r)!} \sum_{i=0}^{r-1}\binom{r-1}{i}\int_0^\infty x^k \left\{1-e^{-x}\right\}^{r-1-i} \\
& \qquad \times \left\{e^{-x}\right\}^{n-r} e^{-x}\,dx \\
& = \frac{1}{2^n}\sum_{i=0}^{r-1}\frac{n!}{(r-1)!\,(n-r)!}\binom{r-1}{i}\int_0^\infty x^k \left\{1-e^{-x}\right\}^{r-1-i} \\
& \qquad \times \left\{e^{-x}\right\}^{n-r} e^{-x}\,dx \\
& + (-1)^k \frac{1}{2^n}\sum_{i=r}^{n}\frac{n!}{(r-1)!\,(n-r)!}\binom{n-r}{i-r}\int_0^\infty x^k \\
& \qquad \times \left\{1-e^{-x}\right\}^{i-r}\left\{e^{-x}\right\}^{r-1} e^{-x}\,dx \\
& = \frac{1}{2^n}\sum_{i=0}^{r-1}\binom{n}{i}\int_0^\infty x^k \frac{(n-i)!}{(r-1-i)!\,(n-r)!}\left\{1-e^{-x}\right\}^{r-1-i} \\
& \qquad \times \left\{e^{-x}\right\}^{n-r} e^{-x}\,dx \\
& + (-1)^k \frac{1}{2^n}\sum_{i=r}^{n}\binom{n}{i}\int_0^\infty x^k \frac{i!}{(i-r)!\,(r-1)!}\left\{1-e^{-x}\right\}^{i-r} \\
& \qquad \times \left\{e^{-x}\right\}^{r-1} e^{-x}\,dx. \qquad (19.25)
\end{aligned}
$$

Noting now that if $X_1, X_2, \ldots, X_m$ is a random sample from a standard exponential $E(1)$ distribution and that $X_{1,m} < X_{2,m} < \cdots < X_{m,m}$ are the corresponding order statistics, then

$$
E\left(X_{\ell,m}^k\right) = \frac{m!}{(\ell-1)!\,(m-\ell)!}\int_0^\infty x^k \left\{1-e^{-x}\right\}^{\ell-1}\left\{e^{-x}\right\}^{m-\ell} e^{-x}\,dx.
$$

$$(19.26)$$

Upon using this formula for the two integrals on the RHS of (19.25), we readily obtain the following relationship:

$$
E\left(V_{r,n}^k\right) = \frac{1}{2^n}\left[\sum_{i=0}^{r-1}\binom{n}{i} E\left(X_{r-i,n-i}^k\right) + (-1)^k \sum_{i=r}^{n}\binom{n}{i} E\left(X_{i-r+1,i}^k\right)\right].
$$

$$(19.27)$$

Thus, the results of Section 18.8 on the standard exponential $E(1)$ order statistics can readily be used to obtain the moments of order statistics from the standard Laplace $L(1)$ distribution by means of (19.27).

Similarly, by considering the product moment $E\left(V_{r,n}\,V_{s,n}\right)$, splitting the double integral over the range $(-\infty < x < y < \infty)$ into three integrals over the range $(-\infty < x < y < 0)$, $(-\infty < x < 0,\ 0 < y < \infty)$, and $(0 < x < y < \infty)$, respectively, and proceeding similarly, we can show that the following relationship holds:

$$
E\left(V_{r,n}\,V_{s,n}\right) \;=\; \frac{1}{2^n}\left[\sum_{i=0}^{r-1}\binom{n}{i}E\left(X_{r-i,n-i}X_{s-i,n-i}\right)\right.
$$

$$
-\sum_{i=r}^{s-1}\binom{n}{i}E\left(X_{i-r+1,i}\right)E\left(X_{s-i,n-i}\right)
$$

$$
\left.+\sum_{i=s}^{n}\binom{n}{i}E\left(X_{i-s+1,i}X_{i-r+1,i}\right)\right]
\qquad (19.28)
$$

for $1 \le r < s \le n$, where, as before, $E\left(X_{k,m}\right)$ and $E\left(X_{k,m}\,X_{\ell,m}\right)$ denote the single and product moments of order statistics from the standard exponential $E(1)$ distribution.

Exercise 19.2 Prove the relation in (19.28).

Remark 19.1 As done by Govindarajulu (1963), the approach above can be generalized to any symmetric distribution. Specifically, if $V_{i,n}$'s denote the order statistics from a distribution symmetric about 0 and $X_{i,n}$'s denote the order statistics from the corresponding folded distribution (folded about 0), then the two relationships above continue to hold.

Exercise 19.3 Let $V_1, V_2, \ldots, V_n$ be a random sample from a distribution $F(x)$ symmetric about 0, and let $V_{1,n} < V_{2,n} < \cdots < V_{n,n}$ be the corresponding order statistics. Further, let $X_{\ell,m}$ denote the ℓth order statistic from a random sample of size m from the corresponding folded distribution with cdf $G(x) = 2F(x) - 1$ (for $x > 0$). Then, prove the following two relationships between the moments of these two sets of order statistics:
For $1 \le r \le n$ and $k \ge 0$,

$$
E\left(V_{r,n}^k\right) = \frac{1}{2^n}\left[\sum_{i=0}^{r-1}\binom{n}{i}E\left(X_{r-i,n-i}^k\right) + (-1)^k\sum_{i=r}^{n}\binom{n}{i}E\left(X_{i-r+1,i}^k\right)\right];
$$

$$
(19.29)
$$

for $1 \leq r < s \leq n$,

$$E\left(V_{r,n}\, V_{s,n}\right) = \frac{1}{2^n}\left[\sum_{i=0}^{r-1}\binom{n}{i}E\left(X_{r-i,n-i}X_{s-i,n-i}\right)\right.$$

$$-\sum_{i=r}^{s-1}\binom{n}{i}E\left(X_{i-r+1,i}\right)E\left(X_{s-i,n-i}\right)$$

$$\left.+\sum_{i=s}^{n}\binom{n}{i}E\left(X_{i-s+1,i}X_{i-r+1,i}\right)\right]. \qquad (19.30)$$

The relationship in (19.27) can also be established by using simple probability arguments as follows. First, for $1 \leq r \leq n$, let us consider the event $V_{r,n} \geq 0$. Given this event, let i ($0 \leq i \leq r-1$) be the number of V's (among $V_1, V_2, \ldots, V_n$) which are negative. Then, since the remaining $(n-i)$ V's (which are nonnegative) form a random sample from the standard exponential $E(1)$ distribution, conditioned on the event $V_{r,n} \geq 0$, we have

$$V_{r,n} \stackrel{d}{=} X_{r-i,n-i} \qquad \text{for } i = 0, 1, \ldots, r-1 \qquad (19.31)$$

with binomial probabilities $\binom{n}{i}\Big/2^n$. Next, let us consider the event $V_{r,n} < 0$. Given this event, let i ($r \leq i \leq n$) be the number of V's (among $V_1, V_2, \ldots, V_n$) which are negative. Then, since the negative of these i V's (which are negative) form a random sample from the standard exponential $E(1)$ distribution, conditioned on the event $V_{r,n} < 0$, we also have

$$V_{r,n} \stackrel{d}{=} -X_{i-r+1,i} \qquad \text{for } i = r, r+1, \ldots, n \qquad (19.32)$$

with binomial probabilities $\binom{n}{i}\Big/2^n$. Combining (19.31) and (19.32), we readily obtain the relation in (19.27).

Exercise 19.4 Using a similar probability argument, prove the relation in (19.28).

CHAPTER 20

GAMMA DISTRIBUTION

20.1 Introduction

In Section 18.9 we discussed the distribution of the sum

$$W_n = X_1 + \cdots + X_n$$

of independent random variables X_k $(k = 1, 2, \ldots, n)$ having the standard exponential $E(1)$ distribution. It was shown that the Laplace transform $\varphi_n(s)$ of W_n is given by

$$\varphi_n(s) = (1 + s)^{-n}, \tag{20.1}$$

and the corresponding pdf $p_n(x)$ is given by

$$p_n(x) = \frac{1}{(n-1)!} \, x^{n-1} e^{-x} \quad \text{if } x > 0. \tag{20.2}$$

It was also shown in Chapter 18 (see Exercise 18.5) that the sum of two positive random variables Y_1 and Y_2 with the common pdf

$$p_{\frac{1}{2}}(x) = \frac{1}{\sqrt{\pi}} \, x^{-1/2} e^{-x}, \qquad x > 0, \tag{20.3}$$

has the standard exponential distribution. In addition, we may recall Eq. (9.22) in which cumulative probabilities of Poisson distributions have been expressed in terms of the pdf in (20.2).

Probability density functions in (20.2) and (20.3) suggest that we consider the family of densities

$$p_\alpha(x) = C(\alpha) x^{\alpha-1} e^{-x}, \qquad x > 0, \tag{20.4}$$

where $C(\alpha)$ depends on α only. It is easy to see that $p_\alpha(x)$ is a pdf if

$$C(\alpha) = \frac{1}{\int_0^\infty x^{\alpha-1} e^{-x} \, dx} = \frac{1}{\Gamma(\alpha)},$$

where $\Gamma(\alpha)$ is the complete gamma function.

20.2   Notations

We say that a random variable X has the *standard gamma distribution* with parameter $\alpha > 0$ if its pdf has the form

$$p_X(x) = \frac{1}{\Gamma(\alpha)} \, x^{\alpha-1} e^{-x} \quad \text{if } x > 0. \tag{20.5}$$

The linear transformation $Y = a + \lambda X$ yields a random variable with pdf

$$p_Y(x) = \frac{1}{\Gamma(\alpha)\lambda^\alpha} \, (x-a)^{\alpha-1} \exp\left\{-\frac{x-a}{\lambda}\right\} \quad \text{if } x > a. \tag{20.6}$$

We will use the notation $Y \sim \Gamma(\alpha, a, \lambda)$ to denote a random variable with pdf (20.6). Hence, $X \sim \Gamma(\alpha, 0, 1)$ corresponds to the standard gamma distribution with pdf (20.5). Note that when $\alpha = 1$, we get the exponential distributions as a subset of the gamma distributions. The special gamma distributions $\Gamma(n/2, 0, 2)$, when n is a positive integer, are called *chi-square distributions* (χ^2 distribution) with n degrees of freedom. These distributions play a very important role in statistical inferential problems.

From (20.5), we have the cdf of $X \sim \Gamma(\alpha, 0, 1)$ (when $\alpha > 1$) as

$$
\begin{aligned}
F_X(x) &= \int_0^x \frac{1}{\Gamma(\alpha)} t^{\alpha-1} e^{-t} \, dt \\
&= -\frac{1}{\Gamma(\alpha)} \int_0^x t^{\alpha-1} \, d(e^{-t}) \\
&= -\frac{1}{\Gamma(\alpha)} x^{\alpha-1} e^{-x} + \frac{1}{\Gamma(\alpha-1)} \int_0^x t^{\alpha-2} e^{-t} \, dt \\
&= -p_X(x) + F_{X'}(x),
\end{aligned}
$$

where X' is a $\Gamma(\alpha-1, 0, 1)$ random variable. Thus, the expression above for the cdf of X presents a recurrence relation in α. Furthermore, if α equals a positive integer n, then repeated integration by parts as above will yield

$$
\begin{aligned}
F_X(x) &= -\frac{1}{\Gamma(n)} x^{n-1} e^{-x} - \cdots - \frac{1}{\Gamma(2)} x e^{-x} + \int_0^x e^{-t} \, dt \\
&= 1 - \sum_{i=0}^{n-1} \frac{e^{-x} x^i}{i!},
\end{aligned}
$$

which is precisely the relationship between the cumulative Poisson probabilities and the gamma distribution noted in Eq. (9.22).

20.3   Mode

Let $X \sim \Gamma(\alpha, 0, 1)$. If $\alpha = 1$ (the case of the exponential distribution), the pdf $p_X(x)$ is a monotonically decreasing function, as we have seen in Chapter 18. If $\alpha > 1$, the gamma $\Gamma(\alpha, 0, 1)$ distribution is unimodal and its mode [the point where density function (20.5) takes the maximal value] is at $x = \alpha - 1$. On the other hand, if $\alpha < 1$, then $p_X(x)$ tends to infinity as $x \to 0$.

20.4 Laplace Transform and Characteristic Function

Let $X \sim \Gamma(\alpha, 0, 1)$. Then, the Laplace transform $\varphi_X(s)$ of X is given by

$$\varphi_X(s) = Ee^{-sX} \quad = \quad \frac{1}{\Gamma(\alpha)} \int_0^\infty x^{\alpha-1} e^{-(1+s)x} \, dx$$

$$= \quad \frac{1}{(1+s)^\alpha} \cdot \tag{20.7}$$

Then the characteristic function $f_X(t) = Ee^{itX}$ has the form

$$f_X(t) = \varphi_X(-it) = \frac{1}{(1-it)^\alpha} \cdot \tag{20.8}$$

As a result, we can write the following expression for the characteristic function of the random variable $Y = a + \lambda X$, which has the general gamma $\Gamma(\alpha, a, \lambda)$ distribution:

$$f_Y(t) = Ee^{itY} = \frac{e^{iat}}{(1-i\lambda t)^\alpha} \cdot \tag{20.9}$$

20.5 Moments

The exponentially decreasing nature of the pdf in (20.5) entails the existence of all the moments of the gamma distribution.

Moments about zero:
   Let $X \sim \Gamma(\alpha, 0, 1)$. Then,

$$a_n = EX^n \quad = \quad \frac{1}{\Gamma(\alpha)} \int_0^\infty x^{n+\alpha-1} e^{-x} \, dx$$

$$= \quad \frac{\Gamma(n+\alpha)}{\Gamma(\alpha)} \,, \qquad n = 1, 2, \ldots. \tag{20.10}$$

In particular, we have

$$EX \quad = \quad \alpha, \tag{20.11}$$
$$EX^2 \quad = \quad \alpha(\alpha+1), \tag{20.12}$$
$$EX^3 \quad = \quad \alpha(\alpha+1)(\alpha+2), \tag{20.13}$$
$$EX^4 \quad = \quad \alpha(\alpha+1)(\alpha+2)(\alpha+3). \tag{20.14}$$

Note that (20.10) is also valid for moments EX^n of negative order $n > -\alpha$.
   Let us now consider $Y \sim \Gamma(\alpha, a, \lambda)$. Since Y can be expressed as $Y = a + \lambda X$, where $X \sim \Gamma(\alpha, 0, 1)$, we can express the moments of Y as

$$EY^n = E(a + \lambda X)^n \quad = \quad \sum_{r=0}^n \binom{n}{r} a^{n-r} \lambda^r EX^r$$

$$= \quad \sum_{r=0}^n \binom{n}{r} a^{n-r} \lambda^r \frac{\Gamma(r+\alpha)}{\Gamma(\alpha)} \cdot \tag{20.15}$$

Central moments:

Central moments of the random variables $Y \sim \Gamma(\alpha, a, \lambda)$ and $V \sim \Gamma(\alpha, 0, \lambda)$ are the same. Now, let X have the standard gamma $\Gamma(\alpha, 0, 1)$ distribution. Then,

$$Y - EY \overset{d}{=} V - EV \overset{d}{=} \lambda(X - EX)$$

and hence

$$
\begin{aligned}
\beta_n &= E(Y - EY)^n = E(V - EV)^n = \lambda^n E(X - EX)^n \\
&= \lambda^n E(X - \alpha)^n = \lambda^n \sum_{r=0}^{n} \binom{n}{r} (-\alpha)^{n-r} EX^r \\
&= \lambda^n n! \sum_{r=0}^{n} \frac{(-\alpha)^{n-r} \Gamma(r + \alpha)}{\Gamma(\alpha) r! \, (n - r)!}, \qquad n = 1, 2, \ldots. \qquad (20.16)
\end{aligned}
$$

As a special case of (20.16), we obtain the variance of the random variable $Y \sim \Gamma(\alpha, a, \lambda)$ as

$$\text{Var } Y = \beta_2 = \alpha \lambda^2. \qquad (20.17)$$

20.6 Shape Characteristics

From (20.16), we get the third central moment of Y to be

$$\beta_3 = 2\alpha \lambda^3$$

using which we readily obtain Pearson's coefficient of skewness as

$$\gamma_1 = \frac{\beta_3}{\beta_2^{3/2}} = \frac{2}{\sqrt{\alpha}}. \qquad (20.18)$$

This reveals that the gamma distribution is positively skewed for all values of the shape parameter α.

Next, we get the fourth central moment of Y from (20.16) to be

$$\beta_4 = (3\alpha^2 + 6\alpha)\lambda^4,$$

using which we readily obtain Pearson's coefficient of kurtosis as

$$\gamma_2 = \frac{\beta_4}{\beta_2^2} = 3 + \frac{6}{\alpha}. \qquad (20.19)$$

This reveals that the gamma distribution is leptokurtic for all values of the shape parameter α.

Furthermore, we note from (20.18) and (20.19) that as $\alpha \to \infty$, γ_1 and γ_2 tend to 0 and 3, respectively (which are the values of skewness and kurtosis of the normal distribution). As we shall see in Section 20.9, the normal distribution is, in fact, the limiting distribution of gamma distributions as the shape parameter $\alpha \to \infty$. Plots of gamma density function presented in Figure 20.1 reveal these properties.

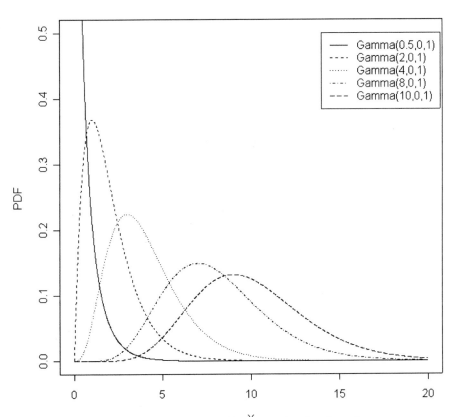

Figure 20.1. Plots of gamma density function

Remark 20.1 From (20.18) and (20.19), we observe that γ_1 and γ_2 satisfy the relationship

$$\gamma_2 = 3 + 1.5\gamma_1^2, \tag{20.20}$$

which is the Type III line in the Pearson plane (that corresponds to the gamma family of distributions).

Exercise 20.1 Generalizing the pdf in (20.5), we may consider the pdf of the *generalized gamma distributions* as

$$p_{X'}(x) = C(\alpha, \delta)x^{\delta\alpha-1}e^{-x^\delta}, \qquad x > 0, \ \alpha > 0, \ \delta > 0.$$

Then, find the normalizing constant $C(\alpha, \delta)$. Derive the moments and discuss the shape characteristics of this generalized gamma family of distributions.

Consider the transformation $Z = \gamma + \log X'$ when X' has a generalized gamma distribution with pdf $p_{X'}(x)$ as given above. Then, we readily obtain the pdf of Z as

$$
\begin{aligned}
p_Z(z) \ = \ & C(\alpha, \delta)e^{\alpha\delta(z-\gamma)}e^{-e^{\delta(z-\gamma)}}, \\
& -\infty < z < \infty, \ \alpha > 0, \ \delta > 0, \ -\infty < \gamma < \infty,
\end{aligned}
$$

and this distribution is called the *log-gamma distribution*. In the above density function, γ is the location parameter, δ is the scale parameter, and α is the shape parameter. The family of log-gamma distributions has found applications in the analysis of life-time data.

Exercise 20.2 For the log-gamma density function presented above, determine the normalizing constant $C(\alpha, \delta)$. Derive the moment generating function of Z and use it to obtain EZ and $\text{Var } Z$.

20.7   Convolutions and Decompositions

Consider independent random variables $Y_1 \sim \Gamma(\alpha, a, \lambda)$ and $Y_2 \sim \Gamma(\beta, b, \lambda)$, having gamma distributions with the same scale parameter $\lambda > 0$. Let $W = Y_1 + Y_2$. Since the characteristic functions of Y_1 and Y_2 are $\dfrac{e^{iat}}{(1 - i\lambda t)^\alpha}$ and $\dfrac{e^{ibt}}{(1 - i\lambda t)^\beta}$, respectively, we obtain the characteristic function $f_W(t)$ of their sum as

$$f_W(t) = \frac{e^{i(a+b)t}}{(1 - i\lambda t)^{\alpha+\beta}} . \tag{20.21}$$

Comparing (20.9) and (20.21), we simply see that $W \sim \Gamma(\alpha + \beta, a + b, \lambda)$. This result is valid for sums of any number of independent random variables $Y_1, Y_2, \ldots, Y_n$, having gamma distributions with the same scale parameter. To be specific, if

$$Y_k \sim \Gamma(\alpha_k, a_k, \lambda), \qquad k = 1, 2, \ldots,$$

then

$$W = Y_1 + Y_2 + \cdots + Y_n \sim \Gamma(\gamma_n, \delta_n, \lambda), \tag{20.22}$$

where $\gamma_n = \alpha_1 + \cdots + \alpha_n$ and $\delta_n = a_1 + \cdots + a_n$.

Moreover, it follows from (20.22) that for any $n = 1, 2, \ldots$, a gamma distributed random variable $W \sim \Gamma(\alpha, a, \lambda)$ can be expressed as a sum of n independent and identically distributed random variables having $\Gamma(\alpha/n, a/n, \lambda)$ distribution. For instance, if $W \sim \Gamma(1, a, \lambda)$ [which means that W has the exponential $E(a, \lambda)$ distribution], then W has the same distribution as that of the sum $Y_1 + Y_2 + \cdots + Y_n$, where $Y_1, Y_2, \ldots, Y_n$ are independent and $Y_k \sim \Gamma(1/n, a/n, \lambda)$, $k = 1, 2, \ldots, n$. This simply means that all gamma distributions (with exponential among them) are infinitely divisible. Also, an example given in Chapter 18 for the exponential distribution (see Exercise 18.5) shows that gamma distributed random variables can admit representations as a sum of non-gamma distributed variables.

20.8   Conditional Distributions and Independence

Let us consider independent random variables $X \sim \Gamma(\alpha, 0, 1)$ and $Y \sim \Gamma(\beta, 0, 1)$. Further, let $V = X + Y$ and $U = X/(X + Y)$. It is then easy to see that the joint density function $p_{X,Y}(x, y)$ of X and Y is given by

$$
\begin{aligned}
p_{X,Y}(x, y) &= p_X(x) p_Y(y) \\
&= \frac{1}{\Gamma(\alpha)\Gamma(\beta)} x^{\alpha-1} y^{\beta-1} e^{-(x+y)}, \quad x > 0, \ y > 0.
\end{aligned}
\tag{20.23}
$$

Taking into account that $X = UV$ and $Y = V(1-U)$ and the Jacobian of the corresponding change of variables $(x, y) = (vu, v(1-u))$ equals v, we obtain from (20.23) the joint density function of the random variables U and V as

$$
\begin{aligned}
p_{U,V}(u, v) &= \frac{1}{\Gamma(\alpha)\Gamma(\beta)} \, v(vu)^{\alpha-1}\{v(1-u)\}^{\beta-1}e^{-v} \\
&= \frac{1}{\Gamma(\alpha)\Gamma(\beta)} \, v^{\alpha+\beta-1}e^{-v}u^{\alpha-1}(1-u)^{\beta-1} \\
&= \frac{1}{\Gamma(\alpha+\beta)} \, v^{\alpha+\beta-1}e^{-v} \frac{1}{B(\alpha+\beta)} \, u^{\alpha-1}(1-u)^{\beta-1},
\end{aligned}
$$
$$ v > 0, \ 0 < u < 1. \qquad (20.24) $$

From (20.24), we observe immediately that the random variables U and V are independent, with V having the gamma $\Gamma(\alpha+\beta, 0, 1)$ distribution (which is to be expected as V is the sum of two independent gamma random variables), and U having the standard beta(α, β) distribution. We need to mention here two interesting special cases. If $X \sim \Gamma\left(\frac{1}{2}, 0, 1\right)$ and $Y \sim \Gamma\left(\frac{1}{2}, 0, 1\right)$, then $U = X/(X + Y)$ has the standard arcsine distribution; and if X and Y have the standard exponential $E(1)$ distribution, then $U = X/(X + Y)$ is a standard uniform $U(0, 1)$ random variable. Of course, we will still have the independence of U and V, and the beta(α, β) distribution of U if we take $X \sim \Gamma(\alpha, 0, \lambda)$ and $Y \sim \Gamma(\beta, 0, \lambda)$.

Lukacs (1965) has proved the converse result: If X and Y are independent positive random variables and the random variables $X/(X + Y)$ and $X + Y$ are also independent, then there exist positive constants α, β, and λ such that $X \sim \Gamma(\alpha, 0, \lambda)$ and $Y \sim \Gamma(\beta, 0, \lambda)$. It was also later shown by Marsaglia (1974) that the result of Lukacs stays true in a more general situation (i.e., without the restriction that X and Y are positive random variables).

Exercise 20.3 Show that the independence of $U = X/(X + Y)$ and $V = X+Y$ also implies the independence of the following pairs of random variables:

(a) $\dfrac{Y}{X}$ and $X + Y$;

(b) $\dfrac{X - Y}{Y}$ and $X + Y$;

(c) $\dfrac{X^2 + Y^2}{XY}$ and $X + Y$;

(d) $\dfrac{(X + Y)^2}{XY}$ and $(X + Y)^2$.

Exercise 20.4 Exploiting the independence of $U = X/(X + Y)$ and $V = X + Y$, the fact that X and Y are both gamma, and the moments of the gamma distribution, derive the moment of the beta distribution in (16.8).

Once again, let $X \sim \Gamma(\alpha, 0, \lambda)$ and $Y \sim \Gamma(\beta, 0, \lambda)$ be independent random variables, and let $V = X + Y$. Consider now the conditional distribution of X given that $V = v$ is fixed. The results presented above enable us to state that conditional distributions of X/v and X/V are the same, given that $V = v$. Thus, the conditional distribution of X/v, given that $X + Y = v$, is beta(α, β). If $\alpha = \beta = 1$ [which means that X and Y have the common exponential $E(\lambda)$ distribution], the conditional distribution of X, given that $X + Y = v$, becomes uniform $U(0, v)$.

The following more general result is also valid for independent gamma random variables. Let $X_k \sim \Gamma(\alpha_k, 0, \lambda)$, $k = 1, 2, \ldots, n$, be independent random variables. Then,

$$\left(\frac{V X_1}{X_1 + \cdots + X_n}, \cdots, \frac{V X_n}{X_1 + \cdots + X_n} \right)$$
$$\stackrel{d}{=} \{X_1, \ldots, X_n \mid V = X_1 + \cdots + X_n = v\}. \qquad (20.25)$$

Recalling now the representation (18.25) for the uniform order statistics $U_{1,n}, \ldots, U_{n,n}$, which has the form

$$\{U_{k,n}\}_{k=1}^{n} \stackrel{d}{=} \left\{ \frac{S_k}{S_{n+1}} \right\}_{k=1}^{n},$$

where $S_k = X_1 + \cdots + X_k$, and

$$X_k \sim \Gamma(1, 0, 1), \qquad k = 1, 2, \ldots$$

are the standard exponential $E(1)$ random variables, we can use (20.25) to obtain another representation for the uniform order statistics as

$$\{U_{1,n}, \ldots, U_{n,n}\} \stackrel{d}{=} \{S_1, \ldots, S_n \mid S_{n+1} = 1\}. \qquad (20.26)$$

20.9   Limiting Distributions

Consider a sequence of random variables $Y_1, Y_2, \ldots$, where

$$Y_n \sim \Gamma(n, 0, 1), \qquad n = 1, 2, \ldots,$$

and with it, generate a new sequence

$$W_n = \frac{Y_n - n}{\sqrt{n}}, \qquad n = 1, 2, \ldots. \qquad (20.27)$$

We then note that

$$W_n \sim \Gamma\left(n, -\sqrt{n}, \frac{1}{\sqrt{n}} \right)$$

with its characteristic function $f_n(t)$ being

$$f_n(t) = \frac{e^{-i\sqrt{n}t}}{(1 - it/\sqrt{n})^n}, \qquad n = 1, 2, \ldots. \qquad (20.28)$$

Exercise 20.5 Show that for any fixed t, as $n \to \infty$,

$$f_n(t) \to e^{-t^2/2}. \tag{20.29}$$

Earlier, we encountered the characteristic function $e^{-t^2/2}$ [for example, in Eqs. (5.43), (7.31) and (9.33)] corresponding to the standard normal distribution. We have, therefore, just established that the normal distribution is the limiting distribution for the sequence of gamma random variables W_n in (20.27).

Exercise 20.6 Let $\Phi(x)$ denote the cumulative distribution function of a random variable with the characteristic function $e^{-t^2/2}$. Later, in Chapter 23, we will find that

$$\Phi(x) = \frac{1}{\sqrt{2\pi}} \int_{-\infty}^{x} e^{-t^2/2} \, dt.$$

Making use of the limiting relation in (20.29), prove that

$$\frac{1}{(n-1)!} \int_{0}^{n+\sqrt{n}} x^{n-1} e^{-x} \, dx \to \Phi(1) = 0.84134\ldots \tag{20.30}$$

as $n \to \infty$.

CHAPTER 21

EXTREME VALUE DISTRIBUTIONS

21.1 Introduction

In Chapter 11 we considered the minimal value $m_n = \min\{U_1, \ldots, U_n\}$ of i.i.d. uniform $U(0,1)$ random variables $U_1, U_2, \ldots$ and determined [see Eq. (11.36)] that the asymptotic distribution of $n\, m_n$ (as $n \to \infty$) becomes the standard exponential distribution. Instead, if we take independent exponential $E(1)$ random variables $X_1, X_2, \ldots$, and generate a sequence of minimal values

$$z_n = \min\{X_1, X_2, \ldots, X_n\}, \qquad n = 1, 2, \ldots,$$

then, as seen in Chapter 18 [see Eq. (18.24)], the sequence $n\, z_n$ converges in distribution (as $n \to \infty$) to the standard exponential distribution.

Consider now the corresponding maximal values

$$M_n = \max\{U_1, \ldots, U_n\}$$

and

$$Z_n = \max\{X_1, X_2, \ldots, X_n\}, \qquad n = 1, 2, \ldots.$$

Then, as seen earlier in Eq. (11.38), as $n \to \infty$,

$$P[n\{M_n - 1\} < x] \to e^x, \qquad x < 0.$$

This simply means that the sequence $n\{1 - M_n\}$ converges asymptotically to the same standard exponential distribution. This fact is not surprising to us since it is clear that the uniform $U(0,1)$ distribution is symmetric with respect to $\frac{1}{2}$ and, consequently,

$$1 - M_n \overset{d}{=} m_n.$$

Let us now find the asymptotic distribution of a suitably normalized maximal value Z_n. We have in this case the following result:

$$
\begin{aligned}
P\{Z_n - \ln n < x\} &= (1 - \exp\{-(x + \ln n)\})^n \\
&= \left(1 - \frac{1}{n} e^{-x}\right)^n \to e^{-e^{-x}}
\end{aligned}
\tag{21.1}
$$

189

as $n \to \infty$. Thus, unlike in the previous cases, we now have a new (non-exponential) distribution for normalized extremes. The natural question that arises here is regarding the set of all possible limiting distributions for maximal and minimal values in a sequence of independent and idetically distributed (i.i.d.) random variables. The evident relationship

$$\max\{-Y_1, -Y_2, \ldots, -Y_n\} \stackrel{d}{=} - \min\{Y_1, Y_2, \ldots, Y_n\} \qquad (21.2)$$

requires us to find only one of these two sets of asymptotic distributions. Indeed, if some cdf $H(x)$ is the limiting distribution for a sequence $\max\{Y_1, Y_2, \ldots, Y_n\}$, $n = 1, 2, \ldots$, then the cdf $G(x) = 1 - H(-x)$ would be the limiting distribution for the sequence $\min\{-Y_1, -Y_2, \ldots, -Y_n\}$, $n = 1, 2, \ldots$, and vice versa.

21.2   Limiting Distributions of Maximal Values

We are interested in all possible limiting cdf's $H(x)$ for sequences

$$H_n(x) = \{F(a_n x + b_n)\}^n, \qquad (21.3)$$

where F is the cdf of the underlying i.i.d. random variables $Y_1, Y_2, \ldots$, and $a_n > 0$ and b_n $(n = 1, 2, \ldots)$ are some normalizing constants; hence, $H_n(x)$ is the cdf of

$$V_n = \frac{\max\{Y_1, Y_2, \ldots, Y_n\} - b_n}{a_n}.$$

Of course, for any F, we can always find a sequence a_n $(n = 1, 2, \ldots)$ which provides the convergence of the sequence V_n to a degenerate limiting distribution. Therefore, our aim here is to find all possible nondegenerate cdf's $H(x)$.

Lemma 21.1 *In order for a nondegenerate cdf $H(x)$ to be the limit of sequence (21.3) for some cdf F and normalizing constants $a_n > 0$ and b_n $(n = 1, 2, \ldots)$, it is necessary and sufficient that for any $s > 0$ and x,*

$$H^s[A(s)x + B(s)] = H(x), \qquad (21.4)$$

where $A(s) > 0$ and $B(s)$ are some functions defined for $s > 0$.

Thus, our problem of finding the asymptotic distributions of maxima is reduced to finding all solutions of the functional equation in (21.4). It turns out that all solutions $H(x)$ (up to location and scale parameters) are as follows:

$$H_1(x) = H_{1,\alpha}(x) = \begin{cases} 0, & x < 0, \\ e^{-x^{-\alpha}}, & x > 0, \ \alpha > 0; \end{cases} \qquad (21.5)$$

$$H_2(x) = H_{2,\alpha}(x) = \begin{cases} e^{-(-x)^\alpha}, & x < 0, \ \alpha > 0, \\ 1, & x > 0; \end{cases} \qquad (21.6)$$

$$H_3(x) = e^{-e^{-x}}, \quad -\infty < x < \infty. \qquad (21.7)$$

Of course, all cdf's

$$H_k\left(\frac{x-a}{\lambda}\right), \qquad k = 1, 2, 3,$$

can also be limiting distributions of normalized maxima for any values of the parameters $-\infty < a < \infty$ and $\lambda > 0$.

H_1 is called the *Fréchet-type distribution*, H_2 the *Weibull-type distribution*, and H_3 the *extreme value distribution*. H_3 is also referred to in the literature as the *log-Weibull, double exponential,* and *doubly exponential distribution*. As mentioned in Chapter 19, the name *double exponential distribution* is sometimes used for the Laplace distribution.

21.3 Limiting Distributions of Minimal Values

As mentioned above, there exists a relation $G(x) = 1 - H(-x)$ between the limiting distributions of maximal $(H(x))$ and minimal $(G(x))$ values. Hence, the set of all possible nondegenerate limiting distributions of the normalized minimal values $\min\{Y_1, Y_2, \ldots, Y_n\}$ are of the following forms:

$$G_1(x) \;=\; G_{1,\alpha}(x) = \begin{cases} 1 - e^{-(-x)^{-\alpha}}, & x < 0, \ \alpha > 0, \\ 1, & x > 0; \end{cases} \qquad (21.8)$$

$$G_2(x) \;=\; G_{2,\alpha}(x) = \begin{cases} 0, & x < 0, \\ 1 - e^{-x^{\alpha}}, & x > 0, \alpha > 0; \end{cases} \qquad (21.9)$$

$$G_3(x) \;=\; 1 - e^{-e^{x}}, \quad -\infty < x < \infty. \qquad (21.10)$$

All cdf's

$$G_k\left(\frac{x-a}{\lambda}\right), \qquad k = 1, 2, 3,$$

where $-\infty < a < \infty$ and $\lambda > 0$, can also be limiting distributions of minimal values. G_2 is commonly known as the *Weibull distribution*.

21.4 Relationships Between Extreme Value Distributions

As seen in the preceding two sections, we have three types of extreme value distributions for maxima and three corresponding types of extreme value distributions for minima. The term *extreme value distributions* includes all distributions with cdf's

$$G_k\left(\frac{x-a}{\lambda}\right) \quad \text{and} \quad H_k\left(\frac{x-a}{\lambda}\right), \qquad k = 1, 2, 3,$$

with the standard members (when $a = 0$ and $\lambda = 1$) being as given in (21.5)–(21.10). Often, the name *extreme value distribution* has been used in the literature only for distributions with cdf's $H_3\left(\dfrac{x-a}{\lambda}\right)$. It is useful to remember that all six types of extreme value distributions given above are closely connected with exponential distributions.

Exercise 21.1 Let X have a standard exponential distribution. Then, show that the random variables

$$-X^{-1/\alpha}, \; X^{1/\alpha}, \; \log X, \; X^{-1/\alpha}, \; -X^{1/\alpha}, \; -\log X$$

have, respectively, the distributions

$$G_{1,\alpha}, \; G_{2,\alpha}, \; G_3, \; H_{1,\alpha}, \; H_{2,\alpha}, \; H_3.$$

Linear transformations of random variables mentioned in Exercise 21.1 enable us to express the distribution of any random variable with cdf's

$$G_k\left(\frac{x-a}{\lambda}\right) \quad \text{and} \quad H_k\left(\frac{x-a}{\lambda}\right), \quad k = 1, 2, 3,$$

via the standard exponential distribution.

Note also that the exponential $E(a, \lambda)$ distribution is a special case of the Weibull distribution because its cdf coincides with $G_{2,1}\left(\dfrac{x-a}{\lambda}\right)$. Furthermore, if we take $Y = \sqrt{X}$, where $X \sim E(1)$, then it is easy to show that

$$F_Y(x) = P\{Y < x\} = 1 - e^{-x^2}, \qquad x > 0. \tag{21.11}$$

We see that the RHS of (21.11) coincides with the Weibull cdf $G_{2,2}(x)$. This distribution of Y is called the *standard Rayleigh distribution*, while linear transformations $a + \lambda Y$ yield the general two-parameter Rayleigh distribution with cdf

$$G_{2,2}\left(\frac{x-a}{\lambda}\right) = 1 - \exp\left\{-\frac{(x-a)^2}{\lambda^2}\right\}, \qquad x > a. \tag{21.12}$$

Exercise 21.2 If X denotes a standard exponential random variable and $Y = \sqrt{X}$, show that the cdf of Y is as given in (21.11). Then, derive the mean and variance of Y.

21.5 Generalized Extreme Value Distributions

It turns out that all the limiting distributions of maxima as well as all the limiting distributions of minima can be presented in a unified form. For this purpose, let us introduce the family of cdf's $H(x, \beta)$ $(-\infty < \beta < \infty)$ which are defined as

$$H(x, \beta) = \exp\{-(1 + x\beta)^{-1/\beta}\} \tag{21.13}$$

in the domain $1 + x\beta > 0$, and we suppose that $H(x, \beta)$ equals 0 or 1 (depending on the sign of β) if $1 + x\beta < 0$. For $\beta = 0$, $H(x, 0)$ means the limit of $H(x, \beta)$ as $\beta \to 0$.

Let us first consider the case $\beta = 1/\alpha$, where $\alpha > 0$. Then,

$$H\left(x, \frac{1}{\alpha}\right) = \begin{cases} \exp\left\{-\left(\dfrac{x+\alpha}{\alpha}\right)^{-\alpha}\right\}, & x > -\alpha, \\ 0, & x < -\alpha. \end{cases} \tag{21.14}$$

It is easy to see that the cdf (21.14) coincides with $H_{1,\alpha}\left(\dfrac{x+\alpha}{\alpha}\right)$.

Next, if $\beta = -1/\alpha$, where $\alpha > 0$, then

$$H\left(x, -\frac{1}{\alpha}\right) = \begin{cases} \exp\left\{-\left(-\dfrac{x-\alpha}{\alpha}\right)^{\alpha}\right\}, & x < \alpha, \\ 1, & x > \alpha. \end{cases} \tag{21.15}$$

This cdf coincides with $H_{2,\alpha}\left(\dfrac{x-\alpha}{\alpha}\right)$.

Finally, for $\beta = 0$, we have

$$H(x, 0) = e^{-e^{-x}} = H_3(x). \tag{21.16}$$

Thus, the derivations above show that the three-parameter family of cdf's

$$H(x, \beta, a, \lambda) = H\left(\frac{x-a}{\lambda}, \beta\right), \quad -\infty < \beta < \infty, \ -\infty < a < \infty, \ \lambda > 0, \tag{21.17}$$

where

$$H(x, \beta) = e^{-(1+x\beta)^{-1/\beta}} \tag{21.18}$$

in the domain $1 + x\beta > 0$, includes all the cdf's

$$H_k\left(\frac{x-a}{\lambda}\right), \qquad k = 1, 2, 3$$

as special cases. Equation (21.17) defines the *generalized extreme value distributions for maxima*, while $H(x, \beta)$ in (21.18) correspond to its standard form. Similarly,

$$G(x, \beta, a, \lambda) = 1 - H(-x, \beta, a, \lambda) \tag{21.19}$$

defines the *generalized extreme value distributions for minima*, and

$$G(x, \beta) = 1 - H(-x, \beta) \tag{21.20}$$

correspond to its standard form.

21.6   Moments

Making use of the representation in Exercise 21.1, we can express moments of the extreme value distributions in terms of moments of the standard exponential distribution. Let random variables Y, W, and V have cdf's $H_{1,\alpha}(x)$, $H_{2,\alpha}(x)$, $H_3(x)$, respectively. Then, from Exercise 21.1, we have the following relations:

$$Y \overset{d}{=} X^{-1/\alpha}, \; W \overset{d}{=} -X^{1/\alpha}, \; V \overset{d}{=} -\log X, \tag{21.21}$$

where X has the standard exponential distribution. Hence, we have

$$EY^k \;\; = \;\; EX^{-k/\alpha} = \int_0^\infty x^{-k/\alpha} e^{-x} \, dx, \tag{21.22}$$

$$EW^k \;\; = \;\; E(-X^{1/\alpha})^k = (-1)^k \int_0^\infty x^{k/\alpha} e^{-x} \, dx, \tag{21.23}$$

and

$$EV^k = E(-\log X)^k = (-1)^k \int_0^\infty \log^k(x) e^{-x} \, dx. \tag{21.24}$$

It readily follows from (21.22) that moments EY^k exist if $k < \alpha$ and that

$$EY^k = \Gamma\left(1 - \frac{k}{\alpha}\right). \tag{21.25}$$

Relation (21.23) reveals that moments EW^k exist for any $\alpha > 0$ and $k = 1, 2, \ldots$, and

$$EW^k = (-1)^k \Gamma\left(1 + \frac{k}{\alpha}\right). \tag{21.26}$$

From (21.25) and (21.26), we also obtain

$$\text{Var } Y = \Gamma\left(1 - \frac{2}{\alpha}\right) - \left\{\Gamma\left(1 - \frac{1}{\alpha}\right)\right\}^2 \tag{21.27}$$

for $\alpha > 2$, and

$$\text{Var } W = \Gamma\left(1 + \frac{2}{\alpha}\right) - \left\{\Gamma\left(1 + \frac{1}{\alpha}\right)\right\}^2 \tag{21.28}$$

for any $\alpha > 0$.

It is known that Euler's constant $\gamma = 0.57722\ldots$ is defined as

$$\gamma = \lim_{n\to\infty}\left(\sum_{k=1}^{n}\frac{1}{k} - \log n\right) \tag{21.29}$$

$$= -\int_{0}^{\infty}\log x\; e^{-x}\; dx. \tag{21.30}$$

Comparing (21.24) and (21.30), we immediately see that

$$EV = \gamma = 0.57722\ldots. \tag{21.31}$$

Another way to obtain (21.31) is through the characteristic function $f_V(t)$ of the random variable V. We have

$$f_V(t) = Ee^{itV} = Ee^{-it\log X} = EX^{-it}$$

$$= \int_{0}^{\infty}x^{-it}e^{-x}\; dx = \Gamma(1-it). \tag{21.32}$$

From (21.32) and the relation

$$f^{(k)}(0) = i^k EV^k\;,$$

we readily find that

$$EV^k = (-1)^k\Gamma^{(k)}(1), \qquad k = 1, 2, \ldots. \tag{21.33}$$

The following useful identity, which is valid for positive z, helps us to find the necessary derivatives of the gamma function:

$$\psi(z) = \frac{d}{dz}\log\Gamma(z) = \frac{\Gamma'(z)}{\Gamma(z)} = \sum_{k=0}^{\infty}\left(\frac{1}{k+1} - \frac{1}{z+k}\right) - \gamma.$$

$$\tag{21.34}$$

Since

$$\Gamma(1) = 1,\;\; \psi(1) = -\gamma \text{ and } \psi'(1) = \sum_{k=0}^{\infty}\frac{1}{(k+1)^2} = \sum_{k=1}^{\infty}\frac{1}{k^2} = \frac{\pi^2}{6}, \tag{21.35}$$

we obtain from (21.33) and (21.34) that

$$EV = -\Gamma'(1) = -\psi(1) = \gamma, \tag{21.36}$$

$$EV^2 = \Gamma^{(2)}(1) = \psi'(1)\Gamma(1) + \psi(1)\Gamma'(1)$$

$$= \psi'(1) + (\psi(1))^2 = \frac{\pi^2}{6} + \gamma^2. \tag{21.37}$$

It follows now from (21.36) and (21.37) that

$$\operatorname{Var} V = \frac{\pi^2}{6}\;. \tag{21.38}$$

Now, let the random variables Y_1, W_1, and V_1 have cdf's $G_{1,\alpha}(x)$, $G_{2,\alpha}(x)$, and $G_3(x)$, respectively. Since

$$Y_1 \overset{d}{=} -Y, \quad W_1 \overset{d}{=} -W, \quad V_1 \overset{d}{=} -V, \tag{21.39}$$

we immediately obtain

$$EY_1^k \;=\; (-1)^k EY^k = (-1)^k \Gamma\left(1 - \frac{k}{\alpha}\right), \quad k < \alpha, \tag{21.40}$$

$$\operatorname{Var} Y_1 \;=\; \operatorname{Var} Y = \Gamma\left(1 - \frac{2}{\alpha}\right) - \left\{\Gamma\left(1 - \frac{1}{\alpha}\right)\right\}^2 \tag{21.41}$$

for $\alpha > 2$,

$$EW_1^k \;=\; (-1)^k EW^k = \Gamma\left(1 + \frac{k}{\alpha}\right), \tag{21.42}$$

$$\operatorname{Var} W_1 \;=\; \operatorname{Var} W = \Gamma\left(1 + \frac{2}{\alpha}\right) - \left\{\Gamma\left(1 + \frac{1}{\alpha}\right)\right\}^2, \tag{21.43}$$

$$EV_1 \;=\; -EV = -\gamma, \tag{21.44}$$

and

$$\operatorname{Var} V_1 = \operatorname{Var} V = \frac{\pi^2}{6}. \tag{21.45}$$

It is important to mention here that extreme value distributions discussed in this chapter have assumed a very important role in life-testing and reliability problems besides being used as probabilistic models in a variety of other problems.

CHAPTER 22

LOGISTIC DISTRIBUTION

22.1 Introduction

Let V_1 and V_2 be i.i.d. random variables having the extreme value distribution with cdf [see (21.7)]

$$H_3(x) = e^{-e^{-x}}, \quad -\infty < x < \infty.$$

Let $V = V_1 - V_2$. Then, the cdf $F_V(x)$ of V is obtained as

$$
\begin{aligned}
F_V(x) &= \int_{-\infty}^{\infty} H_3(x+y)\, dH_3(y) \\
&= \int_{-\infty}^{\infty} e^{-e^{-x-y}} e^{-e^{-y}} e^{-y}\, dy \\
&= \int_{0}^{\infty} e^{-z\left(1+e^{-x}\right)}\, dz \\
&= \frac{1}{1+e^{-x}}, \quad -\infty < x < \infty.
\end{aligned}
\tag{22.1}
$$

This distribution is a particular case of the *logistic distribution*, which has been known since the pioneering work of Verhulst (1838, 1845) on demography.

A book-length account of logistic distributions, discussing in great detail their various properties and applications, is available [Balakrishnan (1992)].

22.2 Notations

A random variable X is said to have a *logistic distribution* if its pdf is given by

$$
p_X(x) = \frac{\pi}{\sigma\sqrt{3}} \frac{e^{-\pi(x-\mu)/(\sigma\sqrt{3})}}{\left\{1 + e^{-\pi(x-\mu)/(\sigma\sqrt{3})}\right\}^2}, \quad -\infty < x < \infty.
\tag{22.2}
$$

The corresponding cdf is given by

$$
F_X(x) = \frac{1}{1 + e^{-\pi(x-\mu)/(\sigma\sqrt{3})}}, \quad -\infty < x < \infty.
\tag{22.3}
$$

We will use $X \sim \text{Lo}(\mu, \sigma^2)$ to denote the random variable X which has the logistic distribution with pdf and cdf as in (22.2) and (22.3), respectively. It is evident that μ $(-\infty < \mu < \infty)$ is the location parameter while σ $(\sigma > 0)$ is the scale parameter. Shortly, we will show that μ and σ^2 are, in fact, the mean and variance of this logistic distribution.

The standard logistic random variable, denoted by $Y \sim \text{Lo}(0, 1)$, has its pdf and cdf as

$$p_Y(x) = \frac{\pi}{\sqrt{3}} \frac{e^{-\pi x/\sqrt{3}}}{\left(1 + e^{-\pi x/\sqrt{3}}\right)^2}, \qquad -\infty < x < \infty \tag{22.4}$$

and

$$F_Y(x) = \frac{1}{1 + e^{-\pi x/\sqrt{3}}}, \qquad -\infty < x < \infty, \tag{22.5}$$

respectively. Although we will see later that this is the standardized form of the distribution (i.e., having zero mean and unit variance), yet it is often convenient to work with the random variable V having the logistic $\text{Lo}\left(0, \pi^2/3\right)$ distribution as it possesses the following simple expressions for the pdf and cdf:

$$p_V(x) = \frac{e^{-x}}{(1 + e^{-x})^2}, \qquad -\infty < x < \infty \tag{22.6}$$

and

$$F_V(x) = \frac{1}{1 + e^{-x}}, \qquad -\infty < x < \infty. \tag{22.7}$$

Since the logistic density function in (22.2) can be rewritten as

$$p_X(x) = \frac{\pi}{4\sigma\sqrt{3}} \text{sech}^2 \left\{ \frac{\pi(x - \mu)}{2\sigma\sqrt{3}} \right\}, \qquad -\infty < x < \infty, \tag{22.8}$$

the logistic distribution is also sometimes referred to as the *sech-squared distribution*.

If $X \sim \text{Lo}(\mu, \sigma^2)$, then we note from (22.2) and (22.3) that the cdf $F_X(x)$ and the pdf $p_X(x)$ satisfy the relationship

$$p_X(x) = \frac{\pi}{\sigma\sqrt{3}} F_X(x) \left\{ 1 - F_X(x) \right\}, \qquad -\infty < x < \infty. \tag{22.9}$$

Of course, in the special case when $V \sim \text{Lo}\left(0, \pi^2/3\right)$, the relationship in (22.9) reduces to

$$p_V(x) = F_V(x) \left\{ 1 - F_V(x) \right\}, \qquad -\infty < x < \infty. \tag{22.10}$$

When $X \sim \text{Lo}(\mu, \sigma^2)$, we may observe from (22.3) that the cdf of X has exponentially decreasing tails. Specifically, we find that

$$F_X(-x) \sim e^{-\pi(x+\mu)/(\sigma\sqrt{3})} \quad \text{and} \quad 1 - F_X(x) \sim e^{-\pi(x-\mu)/(\sigma\sqrt{3})}$$

as $x \to \infty$.

22.3 Moments

Let $V \sim \text{Lo}\left(0, \pi^2/3\right)$. The simple form of its pdf $p_V(x)$ in (22.6) enables us to obtain the moments of V. Then, upon exploiting the obvious relationship

$$X = \frac{\sqrt{3}\,\sigma}{\pi} V + \mu, \qquad (22.11)$$

we can readily obtain the moments of $X \sim \text{Lo}\left(\mu, \sigma^2\right)$. Of course, the exponential rate of decrease of the pdf entails the existence of all the moments of the logistic distribution.

Moments about zero:
  Let $V \sim \text{Lo}\left(0, \pi^2/3\right)$. Since

$$p_V(-x) = \frac{e^x}{\{1 + e^x\}^2} = p_V(x), \qquad -\infty < x < \infty,$$

all the odd-order moments of V are zero. Hence, we need to determine only the even-order moments of V. Now, making use of the identity that

$$(1 + x)^{-2} = \sum_{n=0}^{\infty} \binom{-2}{n} x^n = \sum_{n=0}^{\infty} (-1)^n (n+1) x^n, \qquad (22.12)$$

we derive for $m = 0, 1, 2, \ldots$,

$$
\begin{aligned}
EV^{2m} &= \int_{-\infty}^{\infty} x^{2m} \frac{e^{-x}}{(1 + e^{-x})^2}\, dx \\
&= 2 \int_{0}^{\infty} x^{2m} \frac{e^{-x}}{(1 + e^{-x})^2}\, dx \\
&= 2 \sum_{n=0}^{\infty} (-1)^n (n+1) \int_{0}^{\infty} x^{2m} e^{-(n+1)x}\, dx \\
&= 2\Gamma(2m + 1) \sum_{n=0}^{\infty} \frac{(-1)^n}{(n+1)^{2m}} \\
&= 2(2m)! \sum_{n=1}^{\infty} \frac{(-1)^{n-1}}{n^{2m}} \\
&= 2(2m)! \left\{ \sum_{k=1}^{\infty} \frac{(-1)^{2k-2}}{(2k-1)^{2m}} + \sum_{k=1}^{\infty} \frac{(-1)^{2k-1}}{(2k)^{2m}} \right\} \\
&= 2(2m)! \left\{ \sum_{k=1}^{\infty} (2k-1)^{-2m} - \sum_{k=1}^{\infty} (2k)^{-2m} \right\} \\
&= 2(2m)! \left\{ \sum_{k=1}^{\infty} k^{-2m} - 2 \sum_{k=1}^{\infty} (2k)^{-2m} \right\} \\
&= 2(2m)! \left(1 - 2^{-(2m-1)}\right) \sum_{k=1}^{\infty} k^{-2m} \\
&= 2(2m)! \left(1 - 2^{-(2m-1)}\right) \zeta(2m), \qquad (22.13)
\end{aligned}
$$

where $\zeta(s)$ is the Riemann zeta function defined by

$$\zeta(s) = \sum_{k=1}^{\infty} k^{-s}. \tag{22.14}$$

Using now the well-known facts that $\zeta(2) = \pi^2/6$ and $\zeta(4) = \pi^4/90$, we readily have

$$EV^2 = \operatorname{Var} V = 2\,\zeta(2) = \frac{\pi^2}{3} \tag{22.15}$$

and

$$EV^4 = 42\,\zeta(4) = \frac{7\pi^4}{15}. \tag{22.16}$$

In the general case when $X \sim \text{Lo}\left(\mu, \sigma^2\right)$, using the relationship in (22.11), we readily have

$$
\begin{aligned}
EX^n &= E\left(\frac{\sqrt{3}\,\sigma}{\pi}V + \mu\right)^n \\
&= \sum_{k=0}^{n} \binom{n}{k} \mu^k \left(\frac{\sqrt{3}\,\sigma}{\pi}\right)^{n-k} EV^{n-k}.
\end{aligned} \tag{22.17}
$$

In particular, we find from (22.17) that

$$EX = \mu + \frac{\sqrt{3}\,\sigma}{\pi} EV = \mu \tag{22.18}$$

and

$$EX^2 = \mu^2 + 2\mu\frac{\sqrt{3}\,\sigma}{\pi}EV + \frac{3\sigma^2}{\pi^2}EV^2 = \mu^2 + \sigma^2. \tag{22.19}$$

Central moments:
From (22.11), we find for $n = 0, 1, 2, \ldots,$

$$
\begin{aligned}
\beta_n = E\left(X - EX\right)^n &= \left(\frac{\sqrt{3}\,\sigma}{\pi}\right)^n E\left(V - EV\right)^n \\
&= \left(\frac{\sqrt{3}\,\sigma}{\pi}\right)^n EV^n.
\end{aligned} \tag{22.20}
$$

In particular, we have

$$\beta_2 = \operatorname{Var} X = \frac{3\sigma^2}{\pi^2}EV^2 = \sigma^2 \tag{22.21}$$

and

$$\beta_4 = E\left(X - EX\right)^4 = \frac{9\sigma^4}{\pi^4}EV^4 = \frac{21}{5}\sigma^4 = 4.2\sigma^4. \tag{22.22}$$

Also, due to the symmetry of the distribution of X (about μ), we have

$$\beta_{2n-1} = E\left(X - EX\right)^{2n-1} = E\left(X - \mu\right)^{2n-1} = 0$$
$$\text{for } n = 1, 2, \ldots. \qquad (22.23)$$

22.4 Shape Characteristics

Let X be a logistic $Lo(\mu, \sigma^2)$ random variable. As noted earlier, the distribution of X is symmetric about μ and, consequently, the Pearson coefficient of skewness of X is [see also (22.23)] $\gamma_1 = 0$. In addition, we find from (22.21) and (22.22) the Pearson coefficient of kurtosis of X to be

$$\gamma_2 = \frac{\beta_4}{\beta_2^2} = 4.2. \qquad (22.24)$$

Thus, the logistic distribution is a symmetric leptokurtic distribution. Plots of logistic density function presented in Figure 22.1 (for $\mu = 0$ and different values of σ) reveal these properties.

22.5 Characteristic Function

For simplicity, let us consider $V \sim Lo\left(0, \pi^2/3\right)$ with pdf and cdf as given in (22.6) and (22.7), respectively. Then, the characteristic function of V is determined to be

$$\begin{aligned}
f_V(t) = Ee^{itV} &= \int_{-\infty}^{\infty} \frac{e^{itx-x}}{\left(1 + e^{-x}\right)^2} \, dx \\
&= \int_0^1 u^{it}(1-u)^{-it} \, du \\
&= B(1 + it, 1 - it) \\
&= \Gamma(1 + it)\,\Gamma(1 - it). \qquad (22.25)
\end{aligned}$$

Comparing the characteristic function of $V \sim Lo\left(0, \pi^2/3\right)$ in (22.25) with the characteristic function $\Gamma(1 - it)$ of a random variable V_1 which has the extreme value distribution $H_3(x)$ in (21.7) [see, for example, (21.32)], we readily observe the fact that the logistic $Lo\left(0, \pi^2/3\right)$ random variable V has the same distribution as the difference $V_1 - V_2$, where V_1 and V_2 are i.i.d. random variables with the extreme value distribution $H_3(x)$ in (21.7). Note that this result was derived in Section 22.1 using the convolution formula.

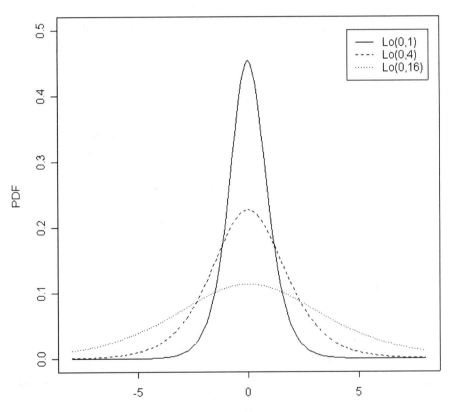

Figure 22.1. Plots of logistic density function when $\mu = 0$

From (22.25), we readily find the characteristic function of $X \sim \text{Lo}(\mu, \sigma^2)$ as

$$f_X(t) = E e^{itX} = e^{it\mu} f_V\left(\frac{\sqrt{3}\,\sigma}{\pi}\,t\right)$$

$$= e^{it\mu}\Gamma\left(1 + \frac{i\sqrt{3}\,\sigma t}{\pi}\right)\Gamma\left(1 - \frac{i\sqrt{3}\,\sigma t}{\pi}\right). \quad (22.26)$$

Now, making use of the facts that

$$\Gamma(z+1) = z\,\Gamma(z) \quad \text{and} \quad \Gamma(z)\Gamma(1-z) = \frac{\pi}{\sin(\pi z)}\;,$$

we can rewrite the characteristic functions in (22.25) and (22.26) as

$$f_V(t) = \Gamma(1+it)\Gamma(1-it) = it\,\Gamma(it)\Gamma(1-it)$$

$$= \frac{\pi it}{\sin(\pi it)} = \frac{\pi t}{\sinh(\pi t)} \quad (22.27)$$

and

$$f_X(t) = e^{it\mu}\frac{i\sqrt{3}\,\sigma t}{\sin(i\sqrt{3}\,\sigma t)} = e^{it\mu}\frac{\sqrt{3}\,\sigma t}{\sinh(\sqrt{3}\sigma t)}\;. \quad (22.28)$$

Exercise 22.1 From the expression of the characteristic function in (22.28), show that the mean and variance of X are μ and σ^2, respectively.

22.6 Relationships with Other Distributions

For simplicity, let us once again consider $V \sim \text{Lo}\left(0, \pi^2/3\right)$ with pdf and cdf as given in (22.6) and (22.7), respectively. Then, using the probability integral transformation, we readily observe the relationship

$$V \stackrel{d}{=} \log\left(\frac{U}{1-U}\right), \quad (22.29)$$

where U has the standard uniform $U(0,1)$ distribution.

Since the distribution of V is symmetric about zero, we may consider the folded form of this logistic distribution, termed the *half logistic distribution* [see, for example, Balakrishnan (1992)]. Specifically, the random variable $|V|$ has the half logistic distribution with pdf and cdf

$$p_{|\nu|}(x) = \frac{2e^{-x}}{(1+e^{-x})^2} \quad \text{and} \quad F_{|\nu|}(x) = \frac{1-e^{-x}}{1+e^{-x}} \quad \text{for } x > 0. \quad (22.30)$$

Exercise 22.2 For the half logistic distribution defined in (22.30), derive the mean and variance.

Realizing that the half logistic distribution in (22.30) is simply the left-truncated (at zero) distribution of $V \sim \text{Lo}\left(0, \pi^2/3\right)$, we can introduce a *general truncated logistic distribution* with pdf

$$p(x) = \frac{e^{-x}}{(B - A)\left(1 + e^{-x}\right)^2}, \quad a < x < b, \tag{22.31}$$

where a and b are the lower and upper points of truncation of the distribution of V, and $A = 1/(1 + e^{-a})$ and $B = 1/(1 + e^{-b})$. Note that the half logistic density function in (22.30) is a special case of the pdf in (22.31) when $a = 0$ and $b = \infty$ so that $A = \frac{1}{2}$ and $B = 1$.

22.7   Decompositions

As already observed, the logistic random variable $V \sim \text{Lo}\left(0, \pi^2/3\right)$ can be represented as the difference $V_1 - V_2$ of two i.i.d. random variables having the extreme value distribution $H_3(x)$ in (21.7). This readily reveals that the logistic random variable is decomposable.

Using Euler's formula on the gamma function given by

$$\Gamma(z) = \int_0^\infty e^{-t} t^{z-1} \, dt = \lim_{n \to \infty} \frac{n!}{z(z+1)(z+2) \cdots (z+n)} n^z, \tag{22.32}$$

we have

$$\begin{aligned} \Gamma(1 + it)\, \Gamma(1 - it) &= \lim_{n \to \infty} \frac{\{(n+1)!\}^2}{(1 + t^2)(4 + t^2) \cdots \{(n+1)^2 + t^2\}} \\ &= \prod_{j=1}^\infty \frac{1}{1 + (t/j)^2}. \end{aligned} \tag{22.33}$$

Recalling that $1/(1 + t^2)$ is the characteristic function of the standard Laplace $L(0,1)$ distribution [see, for example, (19.5)], we obtain the following decomposition result from (22.33):

$$V \overset{d}{=} \prod_{j=1}^\infty \frac{Y_j}{j}, \tag{22.34}$$

where $V \sim \text{Lo}\left(0, \pi^2/3\right)$ and $Y_1, Y_2, \ldots$ are i.i.d. standard Laplace $L(0,1)$ random variables with density function

$$p_Y(x) = \frac{1}{2} e^{-|x|}, \quad -\infty < x < \infty .$$

22.8   Order Statistics

Let $V_1, V_2, \ldots, V_n$ be i.i.d. Lo $\left(0, \pi^2/3\right)$ random variables, and let $V_{1,n} < V_{2,n} < \cdots < V_{n,n}$ denote the corresponding order statistics. Then, the density function of $V_{r,n}$ (for $1 \le r \le n$) is given by

$$p_{V_{r,n}}(x) = \frac{n!}{(r-1)!\,(n-r)!} \left(\frac{1}{1+e^{-x}}\right)^{r-1} \left(\frac{e^{-x}}{1+e^{-x}}\right)^{n-r} \frac{e^{-x}}{(1+e^{-x})^2},$$
$$-\infty < x < \infty. \tag{22.35}$$

From (22.35), we derive the characteristic function of $V_{r,n}$ as

$$f_{V_{r,n}}(t) = Ee^{itV_{r,n}}$$

$$= \frac{n!}{(r-1)!\,(n-r)!} \int_{-\infty}^{\infty} e^{itx} \frac{\left(e^{-x}\right)^{n-r+1}}{(1+e^{-x})^{n+1}}\, dx$$

$$= \frac{n!}{(r-1)!\,(n-r)!} \int_0^1 u^{r+it-1}(1-u)^{n-r+it}\, du$$

$$= \frac{\Gamma(r+it)}{\Gamma(r)} \frac{\Gamma(n-r+1-it)}{\Gamma(n-r+1)}. \tag{22.36}$$

Note that we have $f_{V_{r,n}}(t) = f_{V_{n-r+1,n}}(-t)$ due to the symmetry of the logistic distribution and, consequently,

$$EV_{r,n}^k = (-1)^k EV_{n-r+1,n}^k \tag{22.37}$$

for $1 \le r \le n$ and $k = 1, 2, \ldots$.

Exercise 22.3 From the characteristic function of $V_{r,n}$ in (22.36), derive expressions for the mean and variance of $V_{r,n}$ for $1 \le r \le n$.

22.9   Generalized Logistic Distributions

Due to the simple form of the logistic distribution, several generalizations have been proposed in the literature. Four prominent types of generalized logistic densities are as follows:

The Type I generalized logistic density function is

$$p_V^{\mathrm{I}}(x) = \frac{ae^{-x}}{(1+e^{-x})^{a+1}}, \quad -\infty < x < \infty,\ a > 0; \tag{22.38}$$

The Type II generalized logistic density function is

$$p_V^{\mathrm{II}}(x) = \frac{ae^{-ax}}{(1+e^{-x})^{a+1}}, \quad -\infty < x < \infty,\ a > 0; \tag{22.39}$$

The Type III generalized logistic density function is

$$p_V^{III}(x) = \frac{\Gamma(2a)}{\{\Gamma(a)\}^2} \frac{e^{-ax}}{(1+e^{-x})^{2a}}, \quad -\infty < x < \infty, \; a > 0; \quad (22.40)$$

and the *Type IV generalized logistic density function* is

$$p_V^{IV}(x) = \frac{\Gamma(a+b)}{\Gamma(a)\Gamma(b)} \frac{e^{-bx}}{(1+e^{-x})^{a+b}},$$
$$-\infty < x < \infty, \; a > 0, \; b > 0. \quad (22.41)$$

It should be mentioned that all these forms are special cases of a very general family proposed by Perks (1932).

Exercise 22.4 Show that the characteristic functions of these four generalized logistic distributions are

$$\frac{\Gamma(1-it)\Gamma(a+it)}{\Gamma(a)}, \; \frac{\Gamma(1+it)\Gamma(a+it)}{\Gamma(a)},$$
$$\frac{\Gamma(a-it)\Gamma(a+it)}{\{\Gamma(a)\}^2}, \; \text{and} \; \frac{\Gamma(a+it)\Gamma(b-it)}{\Gamma(a)\Gamma(b)}, \quad (22.42)$$

respectively.

Exercise 22.5 From the characteristic functions in (22.42), derive the expressions of the moments and discuss the shape characteristics.

Exercise 22.6 If V has the Type I generalized logistic density function in (22.38), then prove the following:
   (a) The distribution is negatively skewed for $0 < a < 1$ and positively skewed for $a > 1$;
   (b) The distribution of $-aV$ behaves like standard exponential $E(1)$ when $a \to 0$;
   (c) The distribution of $V - \log a$ behaves like extreme value distribution $H_3(x)$ in (21.7) when $a \to \infty$;
and
   (d) $-V$ has the Type II generalized logistic density function in (22.39).

Exercise 22.7 Let V have the Type I generalized logistic density function in (22.38). Let Y, given τ, have the extreme value density function

$$\tau e^{-x} e^{-\tau e^{-x}},$$

where τ has a gamma distribution with density function

$$\frac{1}{\Gamma(a)} e^{-\tau} \tau^{a-1}, \quad \tau > 0, \; a > 0.$$

Then, show that the marginal density function of Y is the same as that of V.

Exercise 22.8 If V has the Type III generalized logistic density function in (22.40), then prove the following:

(a) The distribution of $\sqrt{2/a}\,V$ behaves like standard normal distribution when $a \to \infty$;

(b) If Y_1 and Y_2 are i.i.d. random variables with *log-gamma density function*

$$p_Y(x|\theta, a) = \frac{\theta^a}{\Gamma(a)} e^{-ax} e^{-\theta e^{-x}}, \quad -\infty < x < \infty, \ a > 0, \ \theta > 0,$$

then the difference $Y_1 - Y_2$ is distributed as V;

and

(c) Let Y_1 and Y_2 be i.i.d. random variables with the distribution of $-\log Z$, where Z given τ is distributed as gamma $\Gamma(\theta\tau, a + b)$ and τ is distributed as beta(a, b). Then, the difference $Y_1 - Y_2$ has the same distribution as V.

Exercise 22.9 If V has the Type IV generalized logistic density function in (22.41), then prove the following:

(a) The distribution is negatively skewed for $a < b$, positively skewed for $a > b$, and is symmetric for $a = b$ (in this case, it simply becomes Type III generalized logistic density function);

(b) If V is distributed as beta(a, b), then $\log(Y/(1 - Y))$ is distributed as V;

and

(c) $-V$ has the Type IV generalized logistic density function in (22.41) with parameters a and b interchanged.

CHAPTER 23

NORMAL DISTRIBUTION

23.1 Introduction

In Chapter 5 [see Eq. (5.39)] we considered a sequence of random variables

$$W_n = \frac{X_n - np}{\sqrt{np(1-p)}} \ , \qquad n = 1, 2, \ldots ,$$

where X_n has binomial $B(n,p)$ distribution and showed that the distribution of W_n converges, as $n \to \infty$, to a limiting distribution with characteristic function

$$f(t) = e^{-t^2/2}. \tag{23.1}$$

Similar result was also obtained for some sequences of negative binomial, Poisson, and gamma distributions [see Eqs. (7.30), (9.33), and (20.27)].
Since

$$\int_{-\infty}^{\infty} |f(t)| \, dt < \infty,$$

the corresponding limiting distribution has a density function $\varphi(x)$, which is given by the inverse Fourier transform

$$\varphi(x) = \frac{1}{2\pi} \int_{-\infty}^{\infty} e^{-itx} f(t) \, dt. \tag{23.2}$$

Since $f(t)$ in (23.1) is an even function, we can express $\varphi(x)$ in (23.2) as

$$\varphi(x) = \frac{1}{2\pi} \int_{-\infty}^{\infty} e^{itx} e^{-t^2/2} \, dt. \tag{23.3}$$

Note that $\varphi(x)$ is differentiable and its first derivative is given by

$$\begin{aligned}
\varphi'(x) &= \frac{i}{2\pi} \int_{-\infty}^{\infty} e^{itx} t e^{-t^2/2} \, dt \\
&= \frac{i}{2\pi} \int_{-\infty}^{\infty} e^{itx} \, d\left(- e^{-t^2/2} \right) \\
&= - \frac{x}{2\pi} \int_{-\infty}^{\infty} e^{itx} e^{-t^2/2} \, dt \\
&= -x \, \varphi(x). \tag{23.4}
\end{aligned}$$

We also have from (23.3) that

$$
\begin{aligned}
\varphi(0) &= \frac{1}{2\pi} \int_{-\infty}^{\infty} e^{-t^2/2} \, dt \\
&= \frac{1}{\pi} \int_{0}^{\infty} e^{-t^2/2} \, dt \\
&= \frac{1}{\pi\sqrt{2}} \int_{0}^{\infty} e^{-u} u^{-1/2} \, du \\
&= \frac{1}{\pi\sqrt{2}} \Gamma\left(\frac{1}{2}\right) \\
&= \frac{1}{\sqrt{2\pi}} \,,
\end{aligned}
\tag{23.5}
$$

since $\Gamma\left(\frac{1}{2}\right) = \sqrt{\pi}$. Upon solving the differential equation in (23.4) using (23.5), we readily obtain

$$
\varphi(x) = \frac{1}{\sqrt{2\pi}} \, e^{-x^2/2}, \quad -\infty < x < \infty, \tag{23.6}
$$

as the pdf corresponding to the characteristic function $f(t) = e^{-t^2/2}$, i.e.,

$$
f(t) = e^{-t^2/2} = \int_{-\infty}^{\infty} e^{itx} \varphi(x) \, dx.
$$

A book-length account of normal distributions, discussing in great detail their various properties and applications, is available [Patel and Read (1997)].

23.2   Notations

We say that a random variable X has the *standard normal distribution* if its pdf is as given in (23.6), and its cdf is given by

$$
\Phi(x) = \int_{-\infty}^{x} \varphi(t) \, dt = \frac{1}{\sqrt{2\pi}} \int_{-\infty}^{x} e^{-t^2/2} \, dt. \tag{23.7}
$$

The linear transformation $Y = a + \sigma X$ $(-\infty < a < \infty, \sigma > 0)$ generates a normal random variable with pdf

$$
\begin{aligned}
p_Y(x) &= p(a, \sigma, x) = \frac{1}{\sigma} \, \varphi\left(\frac{x-a}{\sigma}\right) \\
&= \frac{1}{\sigma\sqrt{2\pi}} \exp\left(-\frac{(x-a)^2}{2\sigma^2}\right), \quad -\infty < x < \infty, \tag{23.8}
\end{aligned}
$$

and cdf

$$
\Phi(a, \sigma, x) = \Phi\left(\frac{x-a}{\sigma}\right) = \frac{1}{\sqrt{2\pi}} \int_{-\infty}^{(x-a)/\sigma} e^{-t^2/2} \, dt. \tag{23.9}
$$

In the sequel, we will use the notation $Y \sim N(a, \sigma^2)$ to denote a random variable Y having the normal distribution with location parameter a and scale parameter $\sigma > 0$. Shortly, we will show that a and σ^2 are, in fact, the mean and variance of Y, respectively. Then, $X \sim N(0, 1)$ will denote that X has the standard normal distribution with pdf and cdf as in (23.6) and (23.7), respectively.

The normal density function first appeared in the papers of de Moivre at the beginning of the eighteenth century as an auxiliary function that approximated binomial probabilities. Some decades later, the normal distribution was given by Gauss and Laplace in the theory of errors and the least squares method, respectively. For this reason, the normal distribution is also sometimes referred to as *Gaussian law, Gauss–Laplace distribution, Gaussian distribution*, and the *second law of Laplace*.

23.3 Mode

It is easy to see that normal $N(a, \sigma^2)$ distribution is unimodal. From (23.8), we see that

$$p'_Y(x) = -\frac{1}{\sigma^3}\sqrt{\frac{2}{\pi}}\,(x - a)\exp\left(-\frac{(x-a)^2}{2\sigma^2}\right),$$

which when equated to 0, yields the mode to be the location parameter a, and the maximal value of the pdf $p(a, \sigma, x)$ is then readily obtained from (23.8) to be $1/(\sigma\sqrt{2\pi})$.

23.4 Entropy

The entropy of a normal distribution possesses an interesting property.

Exercise 23.1 Let $Y \sim N(a, \sigma^2)$. Show that its entropy $H(Y)$ is given by

$$H(Y) = \frac{1}{2} + \log(\sigma\sqrt{2\pi}). \tag{23.10}$$

It is of interest to mention here that among all distributions with fixed mean a and variance σ^2, the maximal value of the entropy is attained for the normal $N(a, \sigma^2)$ distribution.

23.5   Tail Behavior

The normal distribution function has light tails. Let $X \sim N(0,1)$. From Table 23.1, which presents values of the standard normal distribution function $\Phi(x)$, we have

$$
\begin{aligned}
P\{|\xi| > 1\} &= \Phi(-1) + 1 - \Phi(1) = 2\{1 - \Phi(1)\} = 0.3173\ldots, \\
P\{|\xi| > 2\} &= 2\{1 - \Phi(2)\} = 0.0455\ldots, \\
P\{|\xi| > 3\} &= 2\{1 - \Phi(3)\} = 0.0027\ldots, \\
P\{|\xi| > 4\} &= 2\{1 - \Phi(4)\} = 0.000063\ldots.
\end{aligned}
$$

It is easy to obtain that for any $x > 0$,

$$
\begin{aligned}
1 - \Phi(x) &= \frac{1}{\sqrt{2\pi}} \int_x^\infty e^{-t^2/2}\, dt \\
&\leq \frac{1}{x\sqrt{2\pi}} \int_x^\infty t e^{-t^2/2}\, dt \\
&= \frac{1}{x\sqrt{2\pi}} \int_x^\infty d\left(- e^{-t^2/2}\right) \\
&= \frac{1}{x\sqrt{2\pi}} e^{-x^2/2}.
\end{aligned}
\tag{23.11}
$$

Similarly, for any $x > 0$,

$$
\frac{1}{\sqrt{2\pi}} \int_x^\infty \frac{1}{t^2} e^{-t^2/2}\, dt \ \leq \ \frac{1}{x^2\sqrt{2\pi}} \int_x^\infty e^{-t^2/2}\, dt
$$

$$
\leq \frac{1}{x^3\sqrt{2\pi}} e^{-x^2/2}.
\tag{23.12}
$$

Using (23.12), we get the following lower bound for the tail of the normal distribution function:

$$
\begin{aligned}
1 - \Phi(x) &= \frac{1}{\sqrt{2\pi}} \int_x^\infty e^{-t^2/2}\, dt \\
&= \frac{1}{\sqrt{2\pi}} \int_x^\infty \frac{1}{t} d\left(- e^{-t^2/2}\right) \\
&= \frac{1}{x\sqrt{2\pi}} e^{-x^2/2} - \frac{1}{\sqrt{2\pi}} \int_x^\infty \frac{1}{t^2} e^{-t^2/2}\, dt \\
&\geq \frac{1}{x\sqrt{2\pi}} e^{-x^2/2} - \frac{1}{x^3\sqrt{2\pi}} e^{-x^2/2} \\
&= \left(\frac{1}{x} - \frac{1}{x^3}\right) \frac{1}{\sqrt{2\pi}} e^{-x^2/2}.
\end{aligned}
\tag{23.13}
$$

Hence, the asymptotic behavior of $1 - \Phi(x)$ is determined by the inequalities

$$
\left(\frac{1}{x} - \frac{1}{x^3}\right) \frac{1}{\sqrt{2\pi}} e^{-x^2/2} \leq 1 - \Phi(x) \leq \frac{1}{x\sqrt{2\pi}} e^{-x^2/2},
\tag{23.14}
$$

x	$\Phi(x)$	x	$\Phi(x)$	x	$\Phi(x)$	x	$\Phi(x)$
0.00	0.5000	0.88	0.8106	1.76	0.9608	2.64	0.9959
0.02	0.5080	0.90	0.8159	1.78	0.9625	2.66	0.9961
0.04	0.5160	0.92	0.8212	1.80	0.9641	2.68	0.9963
0.06	0.5239	0.94	0.8264	1.82	0.9656	2.70	0.9965
0.08	0.5319	0.96	0.8315	1.84	0.9671	2.72	0.9967
0.10	0.5398	0.98	0.8365	1.86	0.9686	2.74	0.9969
0.12	0.5478	1.00	0.8413	1.88	0.9699	2.76	0.9971
0.14	0.5557	1.02	0.8461	1.90	0.9713	2.78	0.9973
0.16	0.5636	1.04	0.8508	1.92	0.9726	2.80	0.9974
0.18	0.5714	1.06	0.8554	1.94	0.9738	2.82	0.9976
0.20	0.5793	1.08	0.8599	1.96	0.9750	2.84	0.9977
0.22	0.5871	1.10	0.8643	1.98	0.9761	2.86	0.9979
0.24	0.5948	1.12	0.8686	2.00	0.9772	2.88	0.9980
0.26	0.6026	1.14	0.8729	2.02	0.9783	2.90	0.9981
0.28	0.6103	1.16	0.8770	2.04	0.9793	2.92	0.9982
0.30	0.6179	1.18	0.8810	2.06	0.9803	2.94	0.9984
0.32	0.6255	1.20	0.8849	2.08	0.9812	2.96	0.9985
0.34	0.6331	1.22	0.8888	2.10	0.9821	2.98	0.9986
0.36	0.6406	1.24	0.8925	2.12	0.9830	3.00	0.9987
0.38	0.6480	1.26	0.8962	2.14	0.9838	3.02	0.9987
0.40	0.6554	1.28	0.8997	2.16	0.9846	3.04	0.9988
0.42	0.6628	1.30	0.9032	2.18	0.9854	3.06	0.9989
0.44	0.6700	1.32	0.9066	2.20	0.9861	3.08	0.9990
0.46	0.6772	1.34	0.9099	2.22	0.9868	3.10	0.9990
0.48	0.6844	1.36	0.9131	2.24	0.9875	3.12	0.9991
0.50	0.6915	1.38	0.9162	2.26	0.9881	3.14	0.9992
0.52	0.6985	1.40	0.9192	2.28	0.9887	3.16	0.9992
0.54	0.7054	1.42	0.9222	2.30	0.9893	3.18	0.9993
0.56	0.7123	1.44	0.9251	2.32	0.9898	3.20	0.9993
0.58	0.7190	1.46	0.9278	2.34	0.9904	3.22	0.9994
0.60	0.7257	1.48	0.9306	2.36	0.9909	3.24	0.9994
0.62	0.7324	1.50	0.9332	2.38	0.9913	3.26	0.9994
0.64	0.7389	1.52	0.9357	2.40	0.9918	3.28	0.9995
0.66	0.7454	1.54	0.9382	2.42	0.9922	3.30	0.9995
0.68	0.7517	1.56	0.9406	2.44	0.9927	3.32	0.9995
0.70	0.7580	1.58	0.9429	2.46	0.9931	3.34	0.9996
0.72	0.7642	1.60	0.9452	2.48	0.9934	3.36	0.9996
0.74	0.7704	1.62	0.9474	2.50	0.9938	3.38	0.9996
0.76	0.7764	1.64	0.9495	2.52	0.9941	3.40	0.9997
0.78	0.7823	1.66	0.9515	2.54	0.9945	3.42	0.9997
0.80	0.7881	1.68	0.9535	2.56	0.9948	3.44	0.9997
0.82	0.7939	1.70	0.9554	2.58	0.9951	3.46	0.9997
0.84	0.7995	1.72	0.9573	2.60	0.9953	3.48	0.9997
0.86	0.8051	1.74	0.9591	2.62	0.9956	3.50	0.9998

Table 23.1: Standard Normal Cumulative Probabilities

which are valid for any positive x. It readily follows from (23.14) that

$$1 - \Phi(x) \sim \frac{1}{x}\, \varphi(x) \qquad (23.15)$$

as $x \to \infty$, where

$$\varphi(x) = \frac{1}{\sqrt{2\pi}}\, e^{-x^2/2}$$

is the pdf of the standard normal distribution [see (23.6)].

23.6 Characteristic Function

From (23.1), we have the characteristic function of the standard normal $N(0,1)$ distribution to be

$$f(t) = e^{-t^2/2}.$$

If Y has the general $N(a, \sigma^2)$ distribution, we can use the relation $Y = a + \sigma X$, where $X \sim N(0,1)$, in order to obtain the characteristic function of Y as

$$f_Y(t) = Ee^{itY} = e^{iat} f(\sigma t) = \exp\left\{ iat - \frac{\sigma^2 t^2}{2} \right\}. \qquad (23.16)$$

We see that $f_Y(t)$ in (23.16) has the form $\exp\{P_2(t)\}$, where $P_2(t)$ is a polynomial of degree two. Marcinkiewicz (1939) has shown that if a characteristic function $g(t)$ is expressed as $\exp\{P_n(t)\}$, where $P_n(t)$ is a polynomial of degree n, then there are only the following two possibilities:

(a) $n = 1$, in which case $g(t) = e^{iat}$ (degenerate distribution);

(b) $n = 2$, in which case $g(t) = \exp\{iat - \sigma^2 t^2/2\}$ (normal distribution).

In Chapter 12 we presented the definition of stable characteristic functions and stable distributions [see Eq. (12.10)].

Exercise 23.2 Prove that the characteristic function in (23.16) is stable, and hence all normal distributions are stable distributions.

Exercise 23.3 Find the characteristic function of XY when X and Y are independent standard normal random variables.

Exercise 23.4 Let X_1, X_2, X_3, and X_4 be independent standard normal random variables. Then, show that the random variable $Z = X_1 X_2 + X_3 X_4$ has the Laplace distribution.

23.7   Moments

The exponentially decreasing nature of the pdf in (23.8) entails the existence
of all the moments of the normal distribution.

Moments about zero:
  Let $X \sim N(0,1)$. Then, it is easy to see that the pdf $\varphi(x)$ in (23.6) is
symmetric about 0, and hence

$$\alpha_{2n+1} = EX^{2n+1} = 0 \tag{23.17}$$

for $n = 1, 2, \ldots$. In particular, we have from (23.17) that

$$\alpha_1 = EX = 0 \tag{23.18}$$

and

$$\alpha_3 = EX^3 = 0. \tag{23.19}$$

Next, the moments of even order are obtained as follows:

$$
\begin{aligned}
\alpha_{2n} = EX^{2n} &= \frac{1}{\sqrt{2\pi}} \int_{-\infty}^{\infty} x^{2n} e^{-x^2/2} \, dx \\
&= \sqrt{\frac{2}{\pi}} \int_{0}^{\infty} x^{2n} e^{-x^2/2} \, dx \\
&= \sqrt{\frac{2}{\pi}} \, 2^{n-\frac{1}{2}} \int_{0}^{\infty} e^{-u} u^{n-\frac{1}{2}} \, du \\
&= \frac{2^n}{\sqrt{\pi}} \, \Gamma\left(n + \frac{1}{2}\right) \\
&= \frac{2^n}{\sqrt{\pi}} \left(n - \frac{1}{2}\right)\left(n - \frac{3}{2}\right) \cdots \frac{1}{2} \Gamma\left(\frac{1}{2}\right) \\
&= 2^n \left(n - \frac{1}{2}\right)\left(n - \frac{3}{2}\right) \cdots \frac{1}{2} \\
&= \frac{(2n)!}{2^n \, n!}, \quad n = 1, 2, \ldots . \tag{23.20}
\end{aligned}
$$

In particular, we obtain from (23.20) that

$$\alpha_2 = EX^2 = 1 \tag{23.21}$$

and

$$\alpha_4 = EX^4 = 3. \tag{23.22}$$

In general, if $Y = a + \sigma X \sim N(a, \sigma^2)$, we can obtain its moments using
the formula

$$EY^n = E(a + \sigma X)^n = \sum_{r=0}^{n} \binom{n}{r} a^r \sigma^{n-r} EX^{n-r}, \quad n = 1, 2, \ldots, \tag{23.23}$$

which immediately yields

$$EY = a \tag{23.24}$$

and

$$EY^2 = a^2 + \sigma^2. \tag{23.25}$$

Central moments:
Let $X \sim N(0,1)$, $V \sim N(0,\sigma^2)$, and $Y \sim N(a,\sigma^2)$. Then, we have the following relations between central moments of these random variables:

$$
\begin{aligned}
E(Y - EY)^n &= E(V - EV)^n = \sigma^n E(X - EX)^n \\
&= \sigma^n EX^n = \sigma^n \alpha_n, \qquad n = 1, 2, \ldots,
\end{aligned} \tag{23.26}
$$

where α_n are as given in (23.17) and (23.20). In particular, we have

$$\text{Var } Y = \sigma^2 \text{ Var } X = \sigma^2 \alpha_2 = \sigma^2. \tag{23.27}$$

We have thus shown that the location parameter a and the scale parameter σ^2 of normal $N(a,\sigma^2)$ distribution are simply the mean and variance of the distribution, respectively.

Cumulants:
In situations when the logarithm of a characteristic function is simpler to deal with than the characteristic function itself, it is convenient to use the *cumulants*. If $f(t)$ is the characteristic function of the random variable X, then the cumulant γ_k of degree k is defined as follows:

$$\gamma_k = \frac{1}{i^k} \left[\frac{d^k}{dt^k} \log f(t) \right]_{t=0}, \qquad k = 1, 2, \ldots . \tag{23.28}$$

In particular, we have

$$
\begin{aligned}
\gamma_1 &= EX, & (23.29) \\
\gamma_2 &= \text{Var } X, & (23.30) \\
\gamma_3 &= E(X - EX)^3. & (23.31)
\end{aligned}
$$

If the moment EX^k exists, then all cumulants $\gamma_1, \gamma_2, \ldots, \gamma_k$ also exist.
Let us now consider $Y \sim N(a,\sigma^2)$. Since its characteristic function $f_Y(t)$ has the form

$$f_Y(t) = \exp\left\{ iat - \frac{\sigma^2 t^2}{2} \right\},$$

we have

$$\log f_Y(t) = iat - \frac{\sigma^2 t^2}{2} .$$

Hence, we find $\gamma_1 = a$, $\gamma_2 = \sigma^2$, and $\gamma_k = 0$ for $k = 3, 4, \ldots$. Moreover, recalling the Marcinkiewicz's result for characteristic functions mentioned earlier, we obtain the following result: If for some n all cumulants γ_k ($k = n, n+1, \ldots$) of a random variable X are equal to zero, then X has either a degenerate or a normal distribution.

23.8   Shape Characteristics

Let $Y \sim N(a, \sigma^2)$. Then, from (23.26), we readily have

$$E\left(Y - EY\right)^3 = \sigma^3 \alpha_3 = 0, \qquad (23.32)$$

from which we immediately obtain the Pearson coefficient of skewness to be 0.

Next, we have from (23.26) and (23.22) that

$$E\left(Y - EY\right)^4 = \sigma^4 \alpha_4 = 3\sigma^4, \qquad (23.33)$$

from which we immediately obtain the Pearson coefficient of kurtosis to be 3. Thus, we have the normal distributions to be symmetric, unimodal, bell-shaped, and mesokurtic distributions. Plots of normal density function presented in Figure 23.1 (for $a = 0$ and different values of σ) reveal these properties.

Remark 23.1 Recalling now that (see Section 1.4) distributions with coefficient of kurtosis smaller than 3 are classified as platykurtic (light-tailed) and those with larger than 3 are classified as leptokurtic (heavy-tailed), we simply realize that a distribution is considered to be light-tailed or heavy-tailed relative to the normal distribution.

23.9   Convolutions and Decompositions

In order to find the distribution of the sum $Y = Y_1 + Y_2$ of two independent random variables $Y_k \sim N(a_k, \sigma_k^2)$, $k = 1, 2$, we must recall from (23.16) that characteristic functions $f_k(t)$ of these random variables have the form

$$f_k(t) = \exp\left\{ ia_k t - \frac{\sigma_k^2 t^2}{2} \right\}, \qquad k = 1, 2,$$

and hence

$$f_Y(t) = E e^{itY} = f_1(t) f_2(t) = \exp\left\{ iat - \frac{\sigma^2 t^2}{2} \right\},$$

where $a = a_1 + a_2$ and $\sigma^2 = \sigma_1^2 + \sigma_2^2$. This immediately implies that Y has normal $N(a_1 + a_2, \sigma_1^2 + \sigma_2^2)$ distribution. Of course, a more general result is also valid.

Exercise 23.5 For any $n = 1, 2, \ldots$, the sum $Y_1 + Y_2 + \cdots + Y_n$ of independent random variables $Y_k \sim N(a_k, \sigma_k^2)$, $k = 1, 2, \ldots, n$, has normal $N(a, \sigma^2)$ distribution with mean $a = a_1 + \cdots + a_n$ and variance $\sigma^2 = \sigma_1^2 + \cdots + \sigma_n^2$.

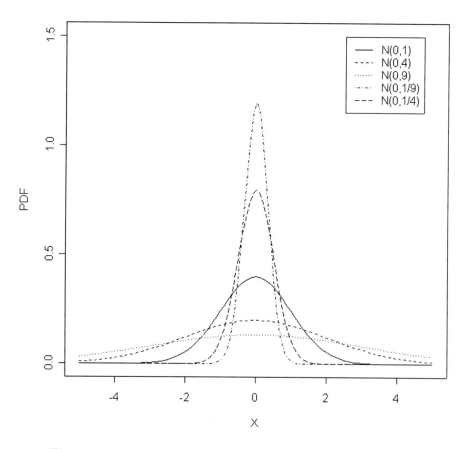

Figure 23.1. Plots of normal density function when $a = 0$

We can now state that any normal $N(a, \sigma^2)$ distribution is decomposable, because it can be presented as a convolution of two normal $N(a_1, \sigma_1^2)$ and $N(a - a_1, \sigma^2 - \sigma_1^2)$ distributions, where $\sigma_1^2 < \sigma^2$ and a_1 may take on any value. In fact, Cramér (1936) has proved that only "normal" decompositions are possible for normal distributions, viz., if $Y \sim N(a, \sigma^2)$, $-\infty < a < \infty$, $\sigma^2 > 0$, and $Y = Y_1 + Y_2$, where Y_1 and Y_2 are independent nondegenerate random variables, then there exist $-\infty < a_1 < \infty$ and $0 < \sigma_1^2 < \sigma^2$ such that $Y_1 \sim N(a_1, \sigma_1^2)$ and $Y_2 \sim N(a - a_1, \sigma^2 - \sigma_1^2)$. Thus, the family of normal distributions is closed with respect to the operations of convolution and decomposition. Recall that families of binomial and Poisson distributions also possess the same property.

Furthermore, the characteristic function of normal $N(a, \sigma^2)$ distribution which is

$$f(t) = \exp\left\{ iat - \frac{\sigma^2 t^2}{2} \right\}$$

can be presented in the form

$$f(t) = \{f_n(t)\}^n,$$

where

$$f_n(t) = \exp\left\{ i\left(\frac{a}{n}\right) t - \frac{\left(\sigma/\sqrt{n}\right)^2 t^2}{2} \right\}$$

is also the characteristic function of a normal distribution. Thus, we have established that any normal distribution is infinitely divisible.

23.10   Conditional Distributions

Consider independent random variables $X \sim N(0, 1)$ and $Y \sim N(0, 1)$. Then, $V = X + Y$ has normal $N(0, 2)$ distribution. Probability density functions of X, Y, and V are given by

$$p_X(x) = p_Y(x) = \varphi(x) = \frac{1}{\sqrt{2\pi}}\, e^{-x^2/2} \tag{23.34}$$

and

$$p_V(x) = \frac{1}{\sqrt{2}}\, \varphi\left(\frac{x}{\sqrt{2}}\right) = \frac{1}{2\sqrt{\pi}}\, e^{-x^2/4}, \tag{23.35}$$

from which we can find the conditional distribution of X given $V = v$. The conditional pdf $p_{X|V}(x|v)$ is given by

$$
\begin{aligned}
p_{X|V}(x|v) &= \frac{p_X(x)\, p_V(v - x)}{p_V(v)} \\
&= \sqrt{2}\, \varphi(x)\, \frac{\varphi(v - x)}{\varphi(v/\sqrt{2})} \\
&= \frac{1}{\sqrt{\pi}} \exp\left\{ -\left(x - \frac{v}{2}\right)^2 \right\} \\
&= \sqrt{2}\varphi\left(\sqrt{2}\left(x - \frac{v}{2}\right)\right).
\end{aligned}
\tag{23.36}
$$

Observing that the RHS of (23.36) is the pdf of normal $N(v/2, 1/2)$ distribution, we conclude that the conditional distribution of X, given $V = v$, is normal with mean $v/2$ and variance $1/2$. A similar result is valid in the general case too.

Exercise 23.6 Find the conditional pdf $p_{X|V}(x|v)$ for the case when $X \sim N(a_1, \sigma_1^2)$, $Y \sim N(a_2, \sigma_2^2)$, with X and Y being independent, and $V = X + Y$.

23.11   Independence of Linear Combinations

As noted earlier (see Exercise 23.3), the sum $\sum_{k=1}^n X_k$ of independent normally distributed random variables $X_k \sim N(a_k, \sigma_k^2)$, $k = 1, 2, \ldots, n$, also has normal

$$N\left(\sum_{k=1}^n a_k, \sum_{k=1}^n \sigma_k^2\right)$$

distribution. The following general result can also be proved similarly.

Exercise 23.7 For any coefficients $b_1, \ldots, b_n$, prove that the linear combination

$$L = \sum_{k=1}^n b_k X_k$$

of independent normally distributed random variables $X_k \sim N(a_k, \sigma_k^2)$, $k = 1, 2, \ldots, n$, also has normal

$$N\left(\sum_{k=1}^n b_k a_k, \sum_{k=1}^n b_k^2 \sigma_k^2\right)$$

distribution.

Let us now consider two different linear combinations

$$L_1 = \sum_{k=1}^n b_k X_k \quad \text{and} \quad L_2 = \sum_{k=1}^n c_k X_k$$

and find the restriction on the coefficients $b_1, \ldots, b_n, c_1, \ldots, c_n$, and parameters $a_1, \ldots, a_n$ and $\sigma_1^2, \ldots, \sigma_n^2$, which provides the independence of L_1 and L_2. It is clear that the location parameters $a_1, \ldots, a_n$ cannot influence this independence, so we will suppose that $a_k = 0$ for $k = 1, 2, \ldots, n$ without loss of any generality. Then, the linear forms L_1 and L_2 are independent if and only if

$$Ee^{iuL_1+ivL_2} = Ee^{iuL_1}Ee^{ivL_2} \qquad (23.37)$$

holds for any real u and v.

Using the fact that $X_k \sim N(0, \sigma_k^2)$, $k = 1, 2, \ldots$, are independent, and that

$$f_k(t) = Ee^{itX_k} = e^{-t^2/2},$$

we get the following expression for the joint characteristic function of L_1 and L_2:

$$
\begin{aligned}
f(u, v) &= Ee^{iuL_1+ivL_2} = Ee^{i\sum_{k=1}^{n}(b_k u+c_k v)X_k} \\
&= \prod_{k=1}^{n} f_k(b_k u + c_k v) \\
&= \exp\left\{ -\frac{1}{2}\sum_{k=1}^{n} \sigma_k^2(b_k u + c_k v)^2 \right\} \\
&= \exp\left\{ -\frac{1}{2}\sum_{k=1}^{n} b_k^2\sigma_k^2 u^2 - \sum_{k=1}^{n} b_k c_k uv\sigma_k^2 - \frac{1}{2}\sum_{k=1}^{n} c_k^2\sigma_k^2 v^2 \right\}.
\end{aligned}
$$
$$(23.38)$$

Also, the characteristic functions of L_1 and L_2 are given by

$$Ee^{iuL_1} = f(u, 0) = \exp\left\{ -\frac{1}{2}\sum_{k=1}^{n} b_k^2\sigma_k^2 u^2 \right\} \qquad (23.39)$$

and

$$Ee^{ivL_2} = f(0, v) = \exp\left\{ -\frac{1}{2}\sum_{k=1}^{n} c_k^2\sigma_k^2 v^2 \right\}. \qquad (23.40)$$

Equations (23.38)–(23.40) then imply that the condition for independence of L_1 and L_2 in (23.37) holds iff

$$\sum_{k=1}^{n} b_k c_k \sigma_k^2 = 0. \qquad (23.41)$$

23.12   Bernstein's Theorem

If we take a pair of i.i.d. normal random variables X and Y and construct linear combinations $L_1 = X + Y$ and $L_2 = X - Y$, then the condition (23.41) is certainly valid and so L_1 and L_2 are independent. Bernstein (1941) proved the converse of this result.

Theorem 23.1 *Let X and Y be i.i.d. random variables, and let $L_1 = X + Y$ and $L_2 = X - Y$ also be independent. Then, X and Y have either degenerate or normal distribution.*

We will present here briefly the main arguments used to prove this theorem.

(1) Indeed, the statement of the theorem is valid if X and Y have degenerate distribution. Hence, we will focus only on the nontrivial situation wherein X and Y have a nondegenerate distribution.

(2) Without loss of generality, we can suppose that X and Y are symmetric random variables with a common nonnegative real characteristic function $f(t)$. This is so because if X and Y have any characteristic function $g(t)$, we can produce symmetric random variables $V = X - X_1$ and $U = Y - Y_1$, where X_1 and Y_1 have the same distribution as X and Y, respectively, and the random variables X, X_1, Y, and Y_1 are all independent. This symmetrization procedure gives us new independent random variables with the common characteristic function $f(t) = g(t)g(-t) = |g(t)|^2$, which is real and nonnegative. Due to the conditions of the theorem, $X + Y$ and $X - Y$ as well as $X_1 + Y_1$ and $X_1 - Y_1$ are independent, and so $X + U$ and $X - U$ are also independent. Thus, the random variables V and U with a common real nonnegative characteristic function satisfy the conditions of the theorem. Suppose now that we have proved that V and U are normally distributed random variables. Since V is the sum of the initial independent random variables X and Y, we can apply Cramér's result on decompositions of normal distributions stated earlier and obtain immediately that X and Y are also normally distributed random variables.

(3) Since X and Y have a common real characteristic function $f(t)$, we have the characteristic functions of L_1 and L_2 as

$$f_1(u) = E e^{iu L_1} = E e^{iu X} E e^{iu Y} = f^2(u) \qquad (23.42)$$

and

$$f_2(v) = E e^{iv L_2} = E e^{iv X} E e^{-iv Y} = f(v)f(-v) = f^2(v). \qquad (23.43)$$

We also have the joint characteristic function of L_1 and L_2 as

$$\begin{aligned} f(u, v) &= E e^{iu L_1 + iv L_2} \\ &= E e^{iu(X+Y)+iv(X-Y)} \\ &= E e^{i(u+v)X+i(u-v)Y} \\ &= E e^{i(u+v)X} E e^{i(u-v)Y} \\ &= f(u+v)f(u-v). \end{aligned} \qquad (23.44)$$

The independence of L_1 and L_2 implies that

$$f(u+v)f(u-v) = f^2(u)f^2(v). \qquad (23.45)$$

(4) Taking $u = nt$ and $v = t$ in (23.45), we get

$$f[(n+1)t]f[(n-1)t] = f^2(nt)f^2(t), \qquad n = 1, 2, \ldots . \qquad (23.46)$$

In particular, we have

$$f(2t) = f^4(t). \qquad (23.47)$$

It follows immediately from (23.47) that $f(t) \neq 0$ for any t. Also, since $f(0) = 1$ and $f(t)$ is a continuous function, there is a value $a > 0$ such that $f(t) \neq 0$ if $|t| < a$. Then, $f(2t) \neq 0$ if $|t| < a$, which means that $f(t) \neq 0$ for all t in the interval $(-2a, 2a)$. Proceeding this way, for any $n = 1, 2, \ldots$, we obtain that $f(t) \neq 0$ if $|t| < 2^n a$, and hence the nonnegative function $f(t)$ must be strictly positive for any t. Now, from the equality [see (23.46) and (23.47)],

$$f(3t)f(t) = f^2(2t)f^2(t) = \{f^4(t)\}^2 f^2(t) = f^{10}(t),$$

we get the relation

$$f(3t) = f^9(t). \qquad (23.48)$$

Following this procedure, we obtain

$$f(mt) = f^{m^2}(t) \qquad (23.49)$$

which is true for any $m = 1, 2, \ldots$ and any t.

(5) The standard technique now allows us to get from (23.49) that

$$f\left(\frac{t}{n}\right) = \{f(t)\}^{1/n^2} \qquad (23.50)$$

and

$$f\left(\frac{m}{n}\right) = c^{(m/n)^2} \qquad (23.51)$$

for any integers m and n, where $c = f(1)$, $0 < c \leq 1$. Since $f(t)$ is a continuous function, (23.51) holds true for any positive t:

$$f(t) = c^{t^2}. \qquad (23.52)$$

Since any real characteristic function is even, (23.52) holds for any t.

If $c = 1$, we obtain $f(t) \equiv 1$ (the characteristic function of the degenerate distribution). If $0 < c < 1$, then

$$f(t) = e^{-\sigma^2 t^2/2},$$

where $\sigma^2 = -2 \log c > 0$, and the theorem is thus proved.

23.13 Darmois–Skitovitch's Theorem

In Bernstein's theorem, we considered the simplest linear combinations $L_1 = X + Y$ and $L_2 = X - Y$. Darmois (1951) and Skitovitch (1954) independently proved the following more general result.

Theorem 23.2 *Let X_k $(k = 1, 2, \ldots, n)$ be independent nondegenerate random variables, and let*

$$L_1 = \sum_{k=1}^{n} b_k X_k \quad \text{and} \quad L_2 = \sum_{k=1}^{n} c_k X_k,$$

where b_k and c_k $(k = 1, 2, \ldots, n)$ are nonzero real coefficients. If L_1 and L_2 are independent, then the random variables $X_1, \ldots, X_n$ all have normal distributions.

A very interesting corollary of Theorem 23.2 is the following. Let

$$L_1 = \bar{X} = \frac{X_1 + \cdots + X_n}{n} \quad \text{and} \quad L_2 = X_1 - \bar{X}.$$

We see that

$$L_1 = \sum_{k=1}^{n} b_k X_k \quad \text{and} \quad L_2 = \sum_{k=1}^{n} c_k X_k,$$

where

$$b_k = \frac{1}{n}, \quad k = 1, 2, \ldots, n,$$

$$c_1 = 1 - \frac{1}{n},$$

$$c_k = -\frac{1}{n}, \quad k = 2, 3, \ldots, n. \tag{23.53}$$

It then follows from Darmois–Skitovitch's theorem that independence of L_1 and L_2 implies that X's are all normally distributed.

If $X_k \sim N(a_k, \sigma_k^2)$, $k = 1, 2, \ldots, n$, then (23.41) shows that for the coefficients in (23.53), linear combinations L_1 and L_2 are independent if and only if

$$\sigma_1^2 = \frac{\sigma_2^2 + \cdots + \sigma_n^2}{n - 1}. \tag{23.54}$$

For example, if $X_k \sim N(a_k, \sigma^2)$, $k = 1, 2, \ldots, n$, then $\sigma_k^2 = \sigma^2$ $(k = 1, 2, \ldots, n)$, in which case (23.54) holds, and so linear combinations L_1 and L_2 are independent.

More generally, in fact, for independent normal random variables $X_k \sim N(a_k, \sigma^2)$, $k = 1, 2, \ldots, n$, the random vector $(X_1 - \bar{X}, X_2 - \bar{X}, \ldots, X_n - \bar{X})$ and $\bar{X} = (X_1 + \cdots + X_n)/n$ are independent. It is clear that in order to prove this result, we can take (without loss of any generality) X's to be standard normal $N(0,1)$ random variables. Let us now consider the joint characteristic function of the random variables $\bar{X}, X_1 - \bar{X}, X_2 - \bar{X}, \ldots, X_n - \bar{X}$:

$$
\begin{aligned}
f(t, t_1, t_2, \ldots, t_n) &= E \exp \left\{ it\bar{X} + i \sum_{k=1}^{n} t_k (X_k - \bar{X}) \right\} \\
&= E \exp \left\{ i \sum_{k=1}^{n} X_k \left(t_k + \frac{t - (t_1 + t_2 + \cdots + t_n)}{n} \right) \right\}.
\end{aligned}
$$

$$(23.55)$$

Since X's in (23.55) are independent random variables having a common characteristic function $f(t) = e^{-t^2/2}$, we obtain

$$
\begin{aligned}
f(t, t_1, t_2, \ldots, t_n) &= \prod_{k=1}^{n} f \left(t_k + \frac{t - (t_1 + t_2 + \cdots + t_n)}{n} \right) \\
&= \prod_{k=1}^{n} \exp \left\{ -\frac{1}{2} \left(t_k + \frac{t - (t_1 + t_2 + \cdots + t_n)}{n} \right)^2 \right\} \\
&= \exp \left\{ -\frac{1}{2} \sum_{k=1}^{n} \left(t_k + \frac{t - (t_1 + t_2 + \cdots + t_n)}{n} \right)^2 \right\}.
\end{aligned}
$$

$$(23.56)$$

We can rewrite (23.56) as

$$
f(t, t_1, t_2, \ldots, t_n) = g(t) h(t_1, t_2, \ldots, t_n), \tag{23.57}
$$

where

$$
g(t) = \exp \left\{ -\frac{t^2}{2n} \right\} \tag{23.58}
$$

and

$$
h(t_1, t_2, \ldots, t_n) = \exp \left\{ -\frac{1}{2} \sum_{k=1}^{n} \left(t_k - \frac{t_1 + t_2 + \cdots + t_n}{n} \right)^2 \right\}. \tag{23.59}
$$

Equation (23.57) then readily implies the independence of $\bar{X}$ and the random vector $(X_1 - \bar{X}, X_2 - \bar{X}, \ldots, X_n - \bar{X})$.

Furthermore, $\bar{X}$ and any random variable $T(X_1 - \bar{X}, \ldots, X_n - \bar{X})$ are also independent. For example, if $X_1, X_2, \ldots, X_n$ is a random sample from normal $N(a, \sigma^2)$ distribution, then we can conclude that the sample mean $\bar{X}$ and the sample variance

$$
S^2 = \frac{1}{n-1} \sum_{k=1}^{n} (X_k - \bar{X})^2 \tag{23.60}
$$

are independent. Note that the converse is also true: If $X_1, X_2, \ldots, X_n$ is a random sample from a distribution and that $\bar{X}$ and S^2 are independent, then X's are all normally distributed.

Now, let

$$X_{1,n} \le X_{2,n} \le \cdots \le X_{n,n}$$

be the order statistics obtained from the random sample $X_1, X_2, \ldots, X_n$. Then, the random vector $(X_{1,n} - \bar{X}, \ldots, X_{n,n} - \bar{X})$ and the sample mean $\bar{X}$ are also independent. This immediately implies, for example, that the sample range

$$X_{n,n} - X_{1,n} = (X_{n,n} - \bar{X}) - (X_{1,n} - \bar{X})$$

and the sample mean $\bar{X}$ are independent for samples from normal distribution.

23.14 Helmert's Transformation

Helmert (1876) used direct transformation of variables to prove, when $X_1, \ldots, X_n$ is a random sample from normal $N(a, \sigma^2)$ distribution, the results that $\bar{X}$ and S^2 are independent and also that $(n-1)S^2/\sigma^2$ has a chi-square distribution with $n-1$ degrees of freedom. Helmert first showed that if $Y_k = X_k - \bar{X}$ $(k = 1, \ldots, n)$, then the joint density function of $Y_1, \ldots, Y_{n-1}$ (with $Y_n = -Y_1 - \ldots - Y_{n-1}$) and $\bar{X}$ is proportional to

$$\exp\left\{-\frac{1}{2\sigma^2}\left(y_1^2 + \cdots + y_n^2\right)\right\} \times \exp\left\{-\frac{n}{2\sigma^2}(\bar{x} - a)^2\right\},$$

thus establishing the independence of $\bar{X}$ and any function of $X_1 - \bar{X}, \ldots, X_n - \bar{X}$, including S^2.

In order to derive the distribution of S^2, Helmert (1876) introduced the transformation

$$w_1 = \sqrt{2}\left(y_1 + \frac{1}{2}y_2 + \frac{1}{2}y_3 + \cdots + \frac{1}{2}y_{n-1}\right),$$

$$w_2 = \sqrt{\frac{3}{2}}\left(y_2 + \frac{1}{3}y_3 + \cdots + \frac{1}{3}y_{n-1}\right),$$

$$w_3 = \sqrt{\frac{4}{3}}\left(y_3 + \frac{1}{4}y_4 + \cdots + \frac{1}{4}y_{n-1}\right),$$

$$\cdots$$

$$w_{n-1} = \sqrt{\frac{n}{n-1}}y_{n-1}.$$

Then, from the joint density function of $Y_1, \ldots, Y_{n-1}$ given by

$$\sqrt{n}\left(\frac{1}{\sqrt{2\pi}\,\sigma}\right)^{n-1}\exp\left\{-\frac{1}{2\sigma^2}\left(y_1^2 + \cdots + y_n^2\right)\right\},$$

we obtain the joint density function of $w_1, \ldots, w_{n-1}$ as

$$\left(\frac{1}{\sqrt{2\pi}\,\sigma}\right)^{n-1}\exp\left\{-\frac{1}{2\sigma^2}\left(w_1^2 + \cdots + w_{n-1}^2\right)\right\}.$$

Since this is the joint density function of $n-1$ independent normal $N(0, \sigma^2)$ random variables, and that

$$\frac{1}{\sigma^2} \sum_{k=1}^{n-1} w_k^2 = \frac{1}{\sigma^2} \sum_{k=1}^{n} y_k^2 = \frac{1}{\sigma^2} \sum_{k=1}^{n} (x_k - \bar{x})^2 = \frac{(n-1)S^2}{\sigma^2},$$

Helmert concluded that the variable $(n-1)S^2/\sigma^2$ has a chi-square distribution with $n-1$ degrees of freedom. The elegant transformation above given is referred to in the literature as *Helmert's transformation*.

23.15   Identity of Distributions of Linear Combinations

Once again, let $X_k \sim N(a_k, \sigma_k^2)$, $k = 1, 2, \ldots, n$, be independent random variables, and

$$L_1 = \sum_{k=1}^{n} b_k X_k \quad \text{and} \quad L_2 = \sum_{k=1}^{n} c_k X_k.$$

It then follows from (23.39) and (23.40) that these linear combinations have the same distribution if

$$\sum_{k=1}^{n} b_k a_k = \sum_{k=1}^{n} c_k a_k \tag{23.61}$$

and

$$\sum_{k=1}^{n} b_k^2 \sigma_k^2 = \sum_{k=1}^{n} c_k^2 \sigma_k^2. \tag{23.62}$$

If X's have a common standard normal distribution, then the condition in (23.62) is equivalent to

$$\sum_{k=1}^{n} b_k^2 = \sum_{k=1}^{n} c_k^2. \tag{23.63}$$

For example, in this situation,

$$\frac{X_1 + \cdots + X_m}{\sqrt{m}} \stackrel{d}{=} \frac{X_1 + \cdots + X_n}{\sqrt{n}} \tag{23.64}$$

for any integers m and n and, in particular,

$$\frac{X_1 + \cdots + X_n}{\sqrt{n}} \stackrel{d}{=} X_1. \tag{23.65}$$

Pólya (1932) showed that if X_1 and X_2 are independent and identically distributed nondegenerate random variables having finite variances, then the equality in distribution of the random variables X_1 and $(X_1 + X_2)/\sqrt{2}$ characterizes the normal distribution. Marcinkiewicz (1939) later proved, under some restrictions on the coefficients b_k and c_k $(k = 1, \ldots, n)$, that if $X_1, \ldots, X_n$ are independent random variables having a common distribution and finite moments of all order, then

$$\sum_{k=1}^{n} b_k X_k \overset{d}{=} \sum_{k=1}^{n} c_k X_k$$

implies that X's all have normal distribution.

23.16   Asymptotic Relations

As already seen in Chapters 5, 7, 9, and 20, the normal distribution arises naturally as a limiting distribution of some sequences of binomial, negative binomial, Poisson, and gamma distributed random variables. A careful look at these situations reveals that the normal distribution has appeared there as a limiting distribution for suitably normalized sums of independent random variables. There are several modifications to the *central limit theorem*, which provide (under different restrictions on the random variables $X_1, X_2, \ldots$) the convergence of sums

$$\frac{S_n - ES_n}{\sqrt{\operatorname{Var} S_n}},$$

where $S_n = X_1 + \cdots + X_n$, to the normal distribution. This is the reason why the normal distributions plays an important role in probability theory and mathematical statistics.

Changing the sum S_n by the maxima $M_n = \max\{X_1, X_2, \ldots, X_n\}$, $n = 1, 2, \ldots$, we get a different limiting scheme with the extreme value distributions (see Chapter 21) determining the asymptotic behavior of the normalized random variable M_n. It should be mentioned here that if $X_k \sim N(0, 1)$, $k = 1, 2, \ldots$, are independent random variables, then

$$P\left\{\frac{M_n - a_n}{b_n} < x\right\} \to e^{-e^{-x}} \tag{23.66}$$

as $n \to \infty$, where

$$a_n = \sqrt{2 \log n} - \frac{\log \log n + \log 4\pi}{2\sqrt{2 \log n}} \tag{23.67}$$

and

$$b_n = \frac{1}{\sqrt{2 \log n}} \,. \tag{23.68}$$

23.17   Transformations

Consider two independent random variables R and ϕ, where ϕ has the uniform $U(0, 2\pi)$ distribution and R has its pdf as

$$p_R(r) = r \, e^{-r^2/2}, \qquad r \geq 0 . \tag{23.69}$$

Note that

$$P\{R < r\} = 1 - e^{-r^2/2}, \qquad r \geq 0, \tag{23.70}$$

which means that R has the *Rayleigh distribution* [see (21.12)], which is a special case of the Weibull distribution. Moreover, R can be expressed as

$$R = \sqrt{2X}, \tag{23.71}$$

where X has the standard exponential $E(1)$ distribution. Then, the joint density function of R and ϕ is given by

$$p_{R,\phi}(r, \varphi) = \frac{1}{2\pi} r \, e^{-r^2/2}, \qquad r \geq 0, \ 0 \leq \varphi \leq 2\pi. \tag{23.72}$$

Let us now consider two new random variables $V = R \sin \phi$ and $W = R \cos \phi$. In fact, R and ϕ are the polar coordinates of the random point (V, W). Since the Jacobian of the polar transformation $(v = r \sin \varphi, \ w = r \cos \varphi)$ equals r, we readily obtain the joint pdf of V and W as

$$p_{V,W}(v, w) = \frac{1}{2\pi} \exp \left\{ -\frac{v^2 + w^2}{2} \right\}, \qquad -\infty < v, w < \infty. \tag{23.73}$$

Equation (23.73) implies that the random variables V and W are independent and both have standard normal distribution. This result, called *Box–Muller's transformation* [Box and Muller (1958)], shows that we have the following representation for a pair of independent random variables having a common standard normal $N(0, 1)$ distribution:

$$(V, W) \overset{d}{=} (\sqrt{2X} \sin 2\pi U, \ \sqrt{2X} \cos 2\pi U), \tag{23.74}$$

where X and U are independent, X has the standard exponential $E(1)$ distribution, and U has the standard uniform $U(0, 1)$ distribution.

We can obtain some interesting corollaries from (23.74). For example,

$$\frac{V^2 + W^2}{2} \overset{d}{=} X \tag{23.75}$$

and so

$$\left(\frac{V}{\sqrt{V^2 + W^2}}, \ \frac{W}{\sqrt{V^2 + W^2}} \right) \overset{d}{=} (\sin 2\pi U, \ \cos 2\pi U). \tag{23.76}$$

We see from (23.75) and (23.76) that the vector on the left-hand side of (23.76) does not depend on $V^2 + W^2$ and it has the uniform distribution on the unit circle. From (23.76), we also have

$$Z = \frac{V}{W} \overset{d}{=} \tan 2\pi U. \tag{23.77}$$

It is easy to see that

$$\tan 2\pi U \stackrel{d}{=} \tan \pi \left(U - \frac{1}{2} \right)$$

and so

$$Z \stackrel{d}{=} \tan \pi \left(U - \frac{1}{2} \right). \tag{23.78}$$

Comparing (23.78) with (12.5), we readily see that Z has the standard Cauchy $C(0,1)$ distribution.

Next, let us consider the random variables

$$Y_1 = \frac{2VW}{\sqrt{V^2 + W^2}} \quad \text{and} \quad Y_2 = \frac{W^2 - V^2}{\sqrt{V^2 + W^2}} . \tag{23.79}$$

Shepp (1964) proved that Y_1 and Y_2 are independent random variables having standard normal distribution. This result can easily be checked by using (23.74). We have

$$
\begin{aligned}
(Y_1, Y_2) &\stackrel{d}{=} \left(2\sqrt{2X} \sin 2\pi U \cos 2\pi U, \; 2\sqrt{2X}(\cos^2 2\pi U - \sin^2 2\pi U) \right) \\
&\stackrel{d}{=} \left(\sqrt{2X} \sin 4\pi U, \; \sqrt{2X} \cos 4\pi U \right). \tag{23.80}
\end{aligned}
$$

Taking into account that

$$(\sin 4\pi U, \; \cos 4\pi U) \stackrel{d}{=} (\sin 2\pi U, \; \cos 2\pi U)$$

and hence

$$(Y_1, Y_2) \stackrel{d}{=} \left(\sqrt{2X} \sin 2\pi U, \; \sqrt{2X} \cos 2\pi U \right),$$

we immediately obtain from (23.74) that

$$(Y_1, Y_2) \stackrel{d}{=} (V, W),$$

which proves Shepp's (1964) result.

Bansal et al. (1999) proved the following converse of Shepp's result. Let V and W be independent and identically distributed random variables, and let Y_1 and Y_2 be as defined in (23.79). If there exist real a and b with $a^2 + b^2 = 1$ such that $aY_1 + bY_2$ has the standard normal distribution, then V and W are standard normal $N(0,1)$ random variables. Now, summarizing the results given above, we have the following characterization of the normal distribution: "Let V and W be independent and identically distributed random variables. Then, $V \sim N(0,1)$ and $W \sim N(0,1)$ iff Y_1 has the standard normal distribution. The same is valid for Y_2 as well."

Coming back to (23.75), we see that $X = (V^2 + W^2)/2$ has the standard exponential $E(1)$ distribution with characteristic function $(1 - it)^{-1}$. Since X is a sum of two independent and identically distributed random variables $V^2/2$

and $W^2/2$, we obtain the characteristic function of $V^2/2$ to be $(1 - it)^{-1/2}$, which corresponds to $\Gamma\left(\frac{1}{2}, 0, 1\right)$ with pdf

$$\frac{1}{\sqrt{\pi\, x}}\, e^{-x}, \qquad x > 0.$$

Then, the squared standard normal variable V^2 has $\Gamma\left(\frac{1}{2}, 0, 2\right)$ with pdf

$$\frac{1}{\sqrt{2\pi x}}\, e^{-x/2}, \qquad x > 0.$$

If we now consider the sum

$$S_n = \sum_{k=1}^{n} V_k^2, \tag{23.81}$$

where $V_1, V_2, \ldots, V_n$ are i.i.d. normal $N(0, 1)$ random variables, then the reproductive property of gamma distributions readily reveals that S_n has $\Gamma\left(n/2, 0, 2\right)$ distribution. In Chapter 20 we mentioned that this special case of $\Gamma\left(n/2, 0, 2\right)$ distributions, where n is an integer, is called a *chi-square* (χ^2) *distribution* with n degrees of freedom. Based on (20.24), the following result is then valid: For any $k = 1, 2, \ldots, n$, the random variables

$$T_{k,n} = \frac{V_k^2}{S_n}$$

and S_n are independent, and $T_{k,n}$ has the standard beta$\left(\dfrac{1}{2}, \dfrac{n-1}{2}\right)$ distribution. In particular, if $n = 2$,

$$T_{1,2} = \frac{V_1^2}{V_1^2 + V_2^2} \qquad \text{and} \qquad T_{2,2} = \frac{V_2^2}{V_1^2 + V_2^2}$$

have the standard arcsine distribution.

Exercise 23.8 Show that the quotient $(V_1^2 + V_2^2)/S_4$ has the standard uniform $U(0, 1)$ distribution.

Exercise 23.9 Show that the pdf of

$$\chi_n = \sqrt{S_n} = \left(\sum_{k=1}^{n} V_k^2\right)^{1/2} \tag{23.82}$$

is given by

$$p_n(x) = \frac{1}{\Gamma\left(n/2\right)\, 2^{(n-2)/2}}\, x^{n-1} e^{-x^2/2}, \qquad x \geq 0. \tag{23.83}$$

The distribution with pdf (23.83) is called as *chi distribution* (χ distribution) with n degrees of freedom. Note that (23.83), when $n = 2$, corresponds to the Rayleigh density in (23.69). The case when $n = 3$ with pdf

$$p_3(x) = \sqrt{\frac{2}{\pi}} \, x^2 e^{-x^2/2}, \qquad x > 0, \tag{23.84}$$

is called the *standard Maxwell distribution*.

Consider now the quotient

$$T = \frac{V}{\sqrt{\sum_{k=1}^{n} V_k^2/n}} = \frac{V}{\sqrt{S_n/n}}, \tag{23.85}$$

where $V, V_1, \ldots, V_n$ are all independent random variables having standard normal distribution. The numerator and denominator of (23.85) are independent and have normal distribution and chi distribution with n degrees of freedom, respectively.

Exercise 23.10 Show that the pdf of T is given by

$$p_T(t) = \frac{\Gamma\left(\dfrac{n+1}{2}\right)}{\sqrt{n\pi}\,\Gamma\left(n/2\right)} \left(1 + \frac{t^2}{n}\right)^{-(n+1)/2}, \qquad -\infty < t < \infty. \tag{23.86}$$

The distribution with pdf (23.86) is called *Student's t distribution with n degrees of freedom*. Note that Student's t distribution with one degree of freedom (i.e., case $n = 1$) is just the standard Cauchy distribution.

Exercise 23.11 As $n \to \infty$, show that the t density in (23.86) converges to the standard normal density function.

Now, let $Y = 1/V^2$ be the reciprocal of a squared standard normal $N(0,1)$ random variable. Then, it can be shown that the pdf of Y is

$$p_Y(x) = \frac{1}{\sqrt{2\pi}} \, x^{-3/2} \exp\left\{-\frac{1}{2x}\right\}, \qquad x > 0. \tag{23.87}$$

It can also be shown that the characteristic function of Y is given by

$$f_Y(t) = E e^{itY} = E e^{it/V^2} = \exp\left\{-|t|^{1/2}\,\frac{1-it}{|t|}\right\}. \tag{23.88}$$

It is not difficult to check that $f_Y(t)$ is a stable characteristic function, and the random variable with pdf (23.87) has a stable distribution. It is of interest to mention here that there are only three stable distributions that possess pdf's in a simple analytical form: normal, Cauchy, and (23.87). Of course, Y as well as any other stable random variable has an infinitely divisible distribution.

We may observe that many distributions related to the normal distribution are infinitely divisible. However, in order to show that not all distributions related closely to normal are infinitely divisible or even decomposable, we will give the following example. It is easy to check that if a random variable X with a characteristic function $f(t)$ and a pdf $p(x)$ has a finite second moment α_2, then $f^*(t) = \dfrac{f^{(2)}(t)}{f^{(2)}(0)}$ is indeed a characteristic function which corresponds to the pdf

$$p^*(x) = \frac{x^2 p(x)}{\alpha_2}\ .$$

Now, let $X \sim N(0,1)$. Then,

$$f(t) = e^{-t^2/2} \quad \text{and} \quad p(x) = \frac{1}{\sqrt{2\pi}}\, e^{-x^2/2};$$

consequently,

$$f^*(t) = (1 - t^2) e^{-t^2/2} \tag{23.89}$$

is the characteristic function of a random variable with pdf

$$p^*(x) = x^2 p(x) = \frac{1}{\sqrt{2\pi}}\, x^2 e^{-x^2/2}. \tag{23.90}$$

The characteristic function in (23.89) and hence the distribution in (23.90) are indecomposable.

Let $X \sim N(0,1)$ and $X = a + b \log Y$. Then, Y is said to have a *lognormal distribution* with parameters a and b. By a simple transformation of random variables, we find the pdf of Y as

$$p_Y(y) = \frac{b}{\sqrt{2\pi}\, y} \exp\left\{ -\frac{1}{2}(a + b \log y)^2 \right\}, \qquad y > 0. \tag{23.91}$$

We may take b to be positive without loss of any generality, since $-X$ has the same distribution as X. An alternative reparametrization is obtained by replacing the parameters a and b by the mean m and standard deviation σ of the random variable $\log Y$. Then, the two sets of parameters satisfy the relationships

$$m = -\frac{a}{b} \quad \text{and} \quad \sigma = \frac{1}{b}, \tag{23.92}$$

so that we have $X = (\log Y - m)/\sigma$, and the lognormal pdf under this reparametrization is given by

$$p_Y(y) = \frac{1}{\sqrt{2\pi}\, \sigma y} \exp\left\{ -\frac{1}{2} \frac{(\log y - m)^2}{\sigma^2} \right\}, \qquad y > 0. \tag{23.93}$$

The lognormal distribution is also sometimes called the *Cobb-Douglas distribution* in the economics literature.

Exercise 23.12 Using the relationship $X = (\log Y - m)/\sigma$, where X is a standard normal variable, show that the kth moment of Y is given by

$$E(Y^k) = E\left(e^{k(m+\sigma X)}\right) = e^{km + \frac{1}{2}k^2\sigma^2}. \qquad (23.94)$$

Then, deduce that

$$EY = e^m \sqrt{\omega} \quad \text{and} \quad \text{Var } Y = e^{2m}\omega(\omega - 1),$$

where $\omega = e^{\sigma^2}$.

Lognormal distributions possess many interesting properties and have also found important applications in diverse fields. A detailed account of these developments on lognormal distributions can be found in the books of Aitchison and Brown (1957) and Crow and Shimizu (1988).

Along the same lines, Johnson (1949) considered the following transformations:

$$X = a + b\log\left(\frac{Y}{1-Y}\right) \quad \text{and} \quad X = a + b\sinh^{-1} Y, \qquad (23.95)$$

where X is once again distributed as standard normal. The distributions of Y in these two cases are called *Johnson's S_B and S_U distributions*, respectively. These distributions have been studied rather extensively in both the statistical and applied literature .

Exercise 23.13 By transformation of variables, derive the densities of Johnson's S_B and S_U distributions.

Exercise 23.14 For the S_U distribution, show that the mean and variance are

$$\sqrt{e^{1/b^2}} \sinh\left(\frac{a}{b}\right) \quad \text{and} \quad \frac{1}{2}\left(e^{1/b^2} - 1\right)\left\{e^{1/b^2}\cosh\left(\frac{2a}{b}\right) + 1\right\},$$

$$(23.96)$$

respectively.

CHAPTER 24

MISCELLANEA

24.1 Introduction

In the preceding thirteen chapters, we have described some of the most important and fundamental continuous distributions. In this chapter we describe briefly some more continuous distributions which play an important role either in statistical inferential problems or in applications.

24.2 Linnik Distribution

In Chapter 19 we found that a random variable V with pdf

$$p_V(x) = \frac{1}{2}\, e^{-x}, \qquad -\infty < x < \infty,$$

has as its characteristic function [see Eq. (19.5)]

$$f_V(t) = \frac{1}{1 + t^2}\,. \tag{24.1}$$

Such a random variable V can be represented as

$$V \stackrel{d}{=} \sqrt{X}\, W, \tag{24.2}$$

where X and W are independent random variables with X having the standard exponential $E(1)$ distribution and W having the normal $N(0, 2)$ distribution with characteristic function

$$g_W(t) = E e^{itW} = e^{-t^2}. \tag{24.3}$$

This result can be established through the use of characteristic functions as follows:

$$
\begin{aligned}
f_V(t) &= E e^{itV} = E e^{it\sqrt{X}W} \\
&= \int_0^\infty e^{-x} E e^{it\sqrt{x}W}\, dx
\end{aligned}
$$

$$= \int_0^\infty e^{-x} g_W(t\sqrt{x})\, dx$$

$$= \int_0^\infty e^{-x-t^2 x}\, dx$$

$$= \frac{1}{1+t^2}\,. \tag{24.4}$$

Thus, from (24.4), we observe that $V = \sqrt{X}\, W$ has the standard Laplace $L(1)$ distribution.

Similarly, let us consider the random variable

$$Y = XZ,$$

where X and Z are independent random variables with X once again having the standard exponential $E(1)$ distribution and Z having the standard Cauchy $C(0,1)$ distribution with characteristic function

$$g_Z(t) = Ee^{itZ} = e^{-|t|}. \tag{24.5}$$

Exercise 24.1 Show that the characteristic function of $Y = XZ$ is

$$f_Y(t) = Ee^{itY} = \frac{1}{1+|t|}\,. \tag{24.6}$$

Note that in the two examples above, $W \sim N(0,2)$ and $Z \sim C(0,1)$ have symmetric stable distributions with characteristic functions of the form $\exp(-|t|^\alpha)$, where $\alpha = 2$ in the first case and $\alpha = 1$ in the second case. Now, more generally, starting with a stable random variable $W(\alpha)$ (for $0 < \alpha \le 2$) with characteristic function

$$g_{W(\alpha)}(t) = e^{-|t|^\alpha}, \tag{24.7}$$

let us consider the random variable

$$Y(\alpha) = X^{1/\alpha} W(\alpha), \tag{24.8}$$

where X and $W(\alpha)$ are independent random variables with X once again having the standard exponential $E(1)$ distribution. Then, we readily obtain the characteristic function of $Y(\alpha)$ as

$$
\begin{aligned}
f_{Y(\alpha)}(t) &= Ee^{itY(\alpha)} = Ee^{itX^{1/\alpha}W(\alpha)} \\
&= \int_0^\infty e^{-x} Ee^{itx^{1/\alpha}W(\alpha)}\, dx \\
&= \int_0^\infty e^{-x} g_{W(\alpha)}(tx^{1/\alpha})\, dx \\
&= \int_0^\infty e^{-x-|t|^\alpha x}\, dx \\
&= \frac{1}{1+|t|^\alpha}\,. \tag{24.9}
\end{aligned}
$$

Thus, we have established in (24.9) that

$$f_{Y(\alpha)}(t) = \frac{1}{1 + |t|^\alpha} \tag{24.10}$$

(for any $0 < \alpha \le 2$) is indeed the characteristic function of a random variable $Y(\alpha)$ which is as defined in (24.8). The distribution with characteristic function (24.10) is called the *Linnik distribution*, since Linnik (1953, 1963) was the first to prove that the RHS of (24.10) is a characteristic function for any $0 < \alpha \le 2$. But this simple proof of the result presented here is due to Devroye (1990).

In addition, from (24.8), we immediately have

$$E\{Y(\alpha)\}^\ell = EX^{\ell/\alpha} E\{W(\alpha)\}^\ell. \tag{24.11}$$

Since

$$EX^{\ell/\alpha} = \Gamma\left(\frac{\ell}{\alpha+1}\right)$$

exists for any $\ell > 0$, the existence of the moment $E\{Y(\alpha)\}^\ell$ is determined by the existence of the moment $E\{W(\alpha)\}^\ell$. Hence, we have

$$E\{|Y(\alpha)|^\ell\} < \infty \tag{24.12}$$

if and only if $0 < \ell < \alpha < 2$, and that (24.12) is true for any $\ell > 0$ when $\alpha = 2$.

24.3 Inverse Gaussian Distribution

The two-parameter inverse Gaussian distribution, denoted by $IG(\mu, \lambda)$, has its pdf as

$$p_X(x) = \sqrt{\frac{\lambda}{2\pi x^3}} \exp\left\{-\frac{\lambda}{2\mu^2 x}(x - \mu)^2\right\}, \qquad x > 0,\ \lambda, \mu > 0, \tag{24.13}$$

and the corresponding cdf as

$$F_X(x) = \Phi\left(\sqrt{\frac{\lambda}{x}}\left(\frac{x}{\mu} - 1\right)\right) + e^{2\lambda/\mu}\Phi\left(-\sqrt{\frac{\lambda}{x}}\left(\frac{x}{\mu} + 1\right)\right),$$
$$x > 0. \tag{24.14}$$

The characteristic function of $IG(\mu, \lambda)$ can be shown to be

$$f_X(t) = Ee^{itX} = \exp\left\{\frac{\lambda}{\mu}\left(1 - \sqrt{1 - \frac{2i\mu^2 t}{\lambda}}\right)\right\}. \tag{24.15}$$

Exercise 24.2 From the characteristic function in (24.15), show that $EX = \mu$ and Var $X = \mu^3/\lambda$.

Exercise 24.3 Show that Pearson coefficients of skewness and kurtosis are given by

$$\gamma_1 = 3\sqrt{\frac{\mu}{\lambda}} \quad \text{and} \quad \gamma_2 = 3 + \frac{15\mu}{\lambda},$$

respectively, thus revealing that $IG(\mu, \lambda)$ distributions are positively skewed and leptokurtic. Note that these distributions are represented by the line $\gamma_2 = 3 + 5\gamma_1^2/3$ in the (γ_1^2, γ_2)-plane.

By taking $\lambda = \mu^2$ in (24.13), we obtain the one-parameter inverse Gaussian distribution with pdf

$$p_X(x) = \frac{\mu}{\sqrt{2\pi x^3}} \exp\left\{-\frac{1}{2x}(x-\mu)^2\right\}, \qquad x > 0, \ \mu > 0, \qquad (24.16)$$

denoted by $IG(\mu, \mu^2)$.

Another one-parameter inverse Gaussian distribution may be derived from (24.13) by letting $\mu \to \infty$. This results in the pdf

$$p_X(x) = \sqrt{\frac{\lambda}{2\pi x^3}} \exp\left(-\frac{\lambda}{2x}\right), \qquad x > 0, \ \lambda > 0.$$

Exercise 24.4 Suppose $X_1, X_2, \ldots, X_n$ are independent inverse Gaussian random variables with X_i distributed as $IG(\mu_i, \lambda_i)$. Then, using the characteristic function in (24.15), show that $\sum_{i=1}^{n} \lambda_i X_i/\mu_i^2$ is distributed as $IG(\mu, \mu^2)$, where $\mu = \sum_{i=1}^{n} \lambda_i/\mu_i$. Show then, when $\mu_i = \mu$ and $\lambda_i = \lambda$ for all $i = 1, 2, \ldots, n$, that the sample mean $\bar{X}$ is distributed as $IG(\mu, n\lambda)$.

Inverse Gaussian distributions have many properties analogous to those of normal distributions. Hence, considerable attention has been paid in the literature to inferential procedures for inverse Gaussian distributions as well as their applications. For a detailed discussion on these developments, one may refer to the books by Chhikara and Folks (1989) and Seshadri (1993, 1998).

24.4   Chi-Square Distribution

In Chapter 20 (see, for example, Section 20.2), we made a passing remark that the special case of gamma $\Gamma\left(n/2, 0, 2\right)$ distribution (where n is a positive integer) is called the *chi-square (χ^2) distribution with n degrees of freedom.* We shall denote the corresponding random variable by χ_n^2. Then, from (20.6), we have its density function as

$$p_{\chi_n^2}(x) = \frac{1}{2^{n/2}\Gamma\left(n/2\right)}\, e^{-x/2}\, x^{(n/2)-1}, \qquad 0 < x < \infty. \qquad (24.17)$$

From (20.9), we also have the characteristic function of χ_n^2 as

$$f_{\chi_n^2}(t) = (1 - 2it)^{-n/2}. \qquad (24.18)$$

From (20.15) and (20.17), we have the mean and variance of χ_n^2 as

$$E\chi_n^2 = n \quad \text{and} \quad \text{Var } \chi_n^2 = 2n. \qquad (24.19)$$

Furthermore, from (20.18) and (20.19), we have the coefficients of skewness and kurtosis of χ_n^2 as

$$\gamma_1 = \frac{2\sqrt{2}}{\sqrt{n}} \quad \text{and} \quad \gamma_2 = 3 + \frac{12}{n}. \qquad (24.20)$$

Also, as shown in Chapter 20, the limiting distribution of the sequence $(\chi_n^2 - n)/\sqrt{2n}$ is standard normal.

Next, let χ_n^2 and χ_m^2 be two independent chi-square random variables, and let $\chi^2 = \chi_n^2 + \chi_m^2$. Then, from (24.18), we obtain the characteristic function of χ^2 as

$$f_{\chi^2}(t) = Ee^{it\chi^2} = Ee^{it\chi_n^2}Ee^{it\chi_m^2} = (1 - 2it)^{-(n+m)/2} \qquad (24.21)$$

which readily implies that χ^2 has a chi-square distribution with $(n+m)$ degrees of freedom. On the other hand, if χ_n^2 and X are independent random variables with X having an arbitrary distribution, and if $\chi^2 = \chi_n^2 + X$ is distributed as chi-square with $(n + m)$ degrees of freedom (where m is a positive integer), then the characteristic function of X is

$$f_X(t) = \frac{f_{\chi^2}(t)}{f_{\chi_n^2}(t)} = \frac{(1 - 2it)^{-(n+m)/2}}{(1 - 2it)^{-n/2}} = (1 - 2it)^{-m/2}, \qquad (24.22)$$

which implies that X is necessarily distributed as χ_m^2.

Let $X_1, \ldots, X_n$ be independent standard normal $N(0, 1)$ random variables. Then, as noted in Chapter 23 [see, for example, Eq. (23.81)],

$$S_n = \sum_{k=1}^{n} X_k^2$$

follows a chi-square distribution with n degrees of freedom. More generally, the following result can be established.

Exercise 24.5 Let $Y_1, \ldots, Y_n$ be a random sample from the normal $N(a, \sigma^2)$ distribution, and $\bar{Y} = \frac{1}{n} \sum_{k=1}^n Y_k$ denote the sample mean. Then, show that [see Eq. (23.60)]

$$\frac{(n-1)S^2}{\sigma^2} = \frac{1}{\sigma^2} \sum_{k=1}^n \left(Y_k - \bar{Y}\right)^2 \overset{d}{=} \chi^2_{n-1} . \qquad (24.23)$$

It is because of this fact that chi-square distributions play a very important role in statistical inferential problems. A book-length account of chi-square distributions, discussing in great detail their various properties and applications, is available [Lancaster (1969)].

24.5 t Distribution

Let $X, X_1, \ldots, X_n$ be i.i.d. random variables having standard normal $N(0, 1)$ distribution. Then, consider the random variable [see also Eq. (23.85)]

$$\frac{X}{\sqrt{\sum_{k=1}^n X_k^2 / n}} = \frac{X}{\sqrt{S_n/n}} , \qquad (24.24)$$

where the numerator and denominator are independent with the numerator having a standard normal distribution and S_n having a chi-square distribution with n degrees of freedom. Then, as given in Exercise 23.8, the pdf of this random variable is given by [see Eq. (23.86)]

$$\frac{\Gamma\left(\dfrac{n+1}{2}\right)}{\sqrt{n\pi}\,\Gamma\left(n/2\right)} \left(1 + \frac{t^2}{n}\right)^{-(n+1)/2} , \qquad -\infty < t < \infty, \qquad (24.25)$$

which is called *Student's t distribution with n degrees of freedom*. Let us denote this distribution by t_n. This is a special form of Karl Pearson's Type VII distribution. Since "Student" (1908) was the first to obtain this result, it is called *Student's distribution*. But sometimes this distribution is called *Fisher's distribution*.

More generally, the following result can be established.

Exercise 24.6 Let $Y_1, \ldots, Y_n$ be a random sample from the normal $N(a, \sigma^2)$ distribution, and $\bar{Y} = \sum_{k=1}^n Y_k / n$ denote the sample mean. Then, with S^2 as defined in (24.23), show that

$$\frac{\sqrt{n}(\bar{Y} - a)}{\sqrt{\sum_{k=1}^n (Y_k - \bar{Y})^2 / (n-1)}} = \frac{\sqrt{n}(\bar{Y} - a)}{S} \overset{d}{=} t_{n-1} . \qquad (24.26)$$

It is for this reason that t distributions play a very important role in statistical inferential problems.

From the density function of $X \overset{d}{=} t_n$ in (24.25), it can be shown that the rth moment of X is finite only for $r < n$. Since the density function is symmetric about $x = 0$, all odd moments of X are zero. If r is even, it can be shown that the rth moment of X (which are also central moments) is given by

$$
\begin{aligned}
EX^r &= n^{r/2} \frac{\Gamma\left(\frac{1}{2}(r+1)\right)\Gamma\left(\frac{1}{2}(n-r)\right)}{\Gamma\left(\frac{1}{2}\right)\Gamma\left(\frac{n}{2}\right)} \\
&= n^{r/2} \frac{1 \cdot 3 \cdots (r-1)}{(n-r)(n-r+2)\cdots(n-2)} .
\end{aligned} \qquad (24.27)
$$

Exercise 24.7 Derive the formula in (24.27).

From the expressions above, we readily obtain the mean, variance, and coefficients of skewness and kurtosis of $X \overset{d}{=} t_n$ as

$$
\begin{aligned}
EX &= 0, \quad \text{Var } X = \frac{n}{n-2} \ (n>2), \\
\gamma_1(X) &= 0 \quad \text{and} \quad \gamma_2(X) = \frac{3(n-2)}{n-4} \ (n>4).
\end{aligned} \qquad (24.28)
$$

It is evident that the t distributions are symmetric, unimodal, bell-shaped and leptokurtic distributions. In addition, as mentioned earlier in Exercise 23.9, t_n distributions converge in limit (as $n \to \infty$) to the standard normal distribution. Plots of the t density function presented in Figures 24.1 and 24.2 reveal these properties.

Exercise 24.8 Let X and Y be i.i.d. random variables with t_n distribution. Then, show that

$$
\frac{\sqrt{n}(Y-X)}{2\sqrt{XY}} \overset{d}{=} t_n, \qquad (24.29)
$$

a result established by Cacoullos (1965).

In a recently published article, Jones (2002) observed that the t_2 distribution has simple forms for its distribution and quantile functions which lead to simple calculations for many properties and measures relating to this distribution. The t density in (24.25) reduces, for the case $n = 2$, simply to

$$p_2(t) = \frac{1}{(2 + t^2)^{3/2}}, \qquad -\infty < t < \infty, \qquad (24.30)$$

from which we readily obtain the cdf as

$$
\begin{aligned}
F_2(t) &= \int_{-\infty}^{t} \frac{1}{(2 + u^2)^{3/2}} \, du \\
&= \int_{-\pi/2}^{\tan^{-1}(t/\sqrt{2})} \frac{1}{2^{3/2}(\sec^2 \theta)^{3/2}} \sqrt{2} \sec^2 \theta \, d\theta \\
&\qquad (\text{setting } u = \sqrt{2} \tan \theta) \\
&= \int_{-\pi/2}^{\tan^{-1}(t/\sqrt{2})} \frac{1}{2} \cos \theta \, d\theta \\
&= \frac{1}{2} \left\{ 1 + \sin \left(\tan^{-1} \left(\frac{t}{\sqrt{2}} \right) \right) \right\} \\
&= \frac{1}{2} \left\{ 1 + \sqrt{1 - \frac{1}{1 + \tan^2 \left(\tan^{-1} \left(\frac{t}{\sqrt{2}} \right) \right)}} \right\} \\
&= \frac{1}{2} \left\{ 1 + \frac{t}{\sqrt{2 + t^2}} \right\}, \qquad -\infty < t < \infty. \qquad (24.31)
\end{aligned}
$$

Exercise 24.9 From (24.31), show that the quantile function of the t_2 distribution is

$$F_2^{-1}(u) = \frac{2u - 1}{\sqrt{2u(1 - u)}}, \qquad 0 < u < 1.$$

Proceeding similarly, the following expressions can be obtained corresponding to $n = 3$, $n = 4$, $n = 5$, and $n = 6$:

$$
\begin{aligned}
F_3(t) &= \frac{1}{2} + \frac{1}{\pi} \tan^{-1} \left(\frac{t}{\sqrt{3}} \right) + \frac{\sqrt{3}\, t}{\pi(3 + t^2)}, \\
F_4(t) &= \frac{1}{2} + \frac{t(6 + t^2)}{2(4 + t^2)^{3/2}}, \\
F_5(t) &= \frac{1}{2} + \frac{1}{\pi} \tan^{-1} \left(\frac{t}{\sqrt{5}} \right) + \frac{\sqrt{5}\, t}{\pi(5 + t^2)} + \frac{10\sqrt{5}\, t}{3\pi(5 + t^2)^2}, \\
F_6(t) &= \frac{1}{2} + \frac{t(135 + 30t^2 + 2t^4)}{4(6 + t^2)^{5/2}}.
\end{aligned}
$$

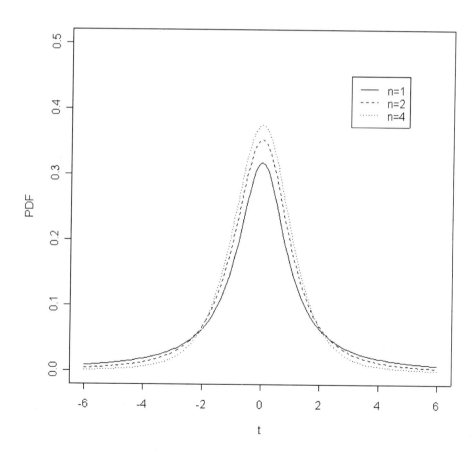

Figure 24.1. Plots of *t* density function when $n = 1, 2, 4$

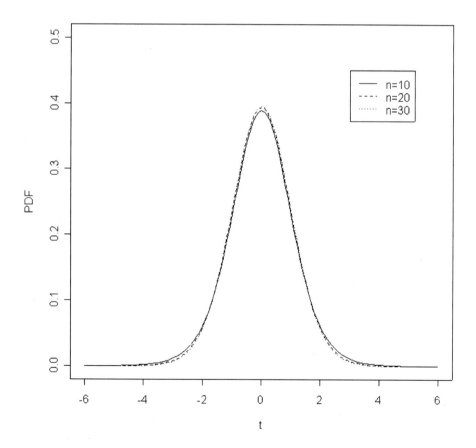

Figure 24.2. Plots of t density function when $n = 10, 20, 30$

24.6 F Distribution

Let $V \sim \chi_m^2$ and $W \sim \chi_n^2$ be independent random variables. Further, let

$$X = \frac{V/m}{W/n} = \frac{n}{m} \cdot \frac{V}{W} . \qquad (24.32)$$

Then, it can be shown that the pdf of X is given by

$$p_X(x) = \frac{(m/n)^{m/2}}{B(m/2, n/2)} \cdot \frac{x^{(m/2)-1}}{(1 + mx/n)^{(m+n)/2}}, \quad x > 0. \qquad (24.33)$$

This is called *Snedecor's F distribution with* (m, n) *degrees of freedom*, as its original derivation is due to Snedecor (1934), and we shall denote it by $F_{m,n}$. This is related to Karl Pearson's Type VI distribution, which is a beta distribution of the second kind mentioned earlier (see Section 16.4).

Exercise 24.10 Derive the density function in (24.33).

From (24.32), we obtain the rth moment of $X \sim F_{m,n}$ as

$$\begin{aligned}
EX^r &= \left(\frac{n}{m}\right)^r E(V^r) E(W^{-r}) \\
&= \left(\frac{n}{m}\right)^r \frac{m(m+2)\cdots(m+2r-2)}{(n-2)(n-4)\cdots(n-2r)} \qquad (24.34)
\end{aligned}$$

for $r < n/2$. From (24.34), we immediately obtain the mean and variance of $X \sim F_{m,n}$ as

$$EX = \frac{n}{n-2} \ (n > 2) \text{ and Var } X = \frac{2n^2(m+n-2)}{m(n-2)^2(n-4)} \ (n > 4). \qquad (24.35)$$

Exercise 24.11 If $X \overset{d}{=} F_{n,n}$, then show that $\sqrt{n}\left(\sqrt{X} - 1/\sqrt{X}\right)/2 \overset{d}{=} t_n$.

24.7   Noncentral Distributions

In Section 24.4 we noted that when $X_1, X_2, \ldots, X_n$ are i.i.d. $N(0,1)$ random variables, the variable $S_n = \sum_{k=1}^{n} X_k^2$ has a chi-square distribution with n degrees of freedom. Now, consider the distribution of the variable

$$S_n' = \sum_{k=1}^{n} (X_k + a_k)^2. \qquad (24.36)$$

The distribution of S_n' depends on $a_1, a_2, \ldots, a_n$ only through $\lambda = \sum_{k=1}^{n} a_k^2$, and is called the *noncentral chi-square distribution* with n degrees of freedom and noncentality parameter $\lambda = \sum_{k=1}^{n} a_k^2$. When $\lambda = 0$ (i.e., when all $a_1, \ldots, a_n$ are zero), this noncentral chi-square distribution becomes the (central) chi-square distribution in (24.17).

Exercise 24.12 Let $Y_1, Y_2, \ldots, Y_n$ be independent random variables with Y_k distributed as $N(a_k, \sigma^2)$, and $\bar{Y} = \sum_{k=1}^{n} Y_k/n$ denote the sample mean. Then, show that

$$\frac{(n-1)S^2}{\sigma^2} = \frac{1}{\sigma^2} \sum_{k=1}^{n} (Y_k - \bar{Y})^2$$

is distributed as noncentral chi-square with $n - 1$ degrees of freedom and noncentrality parameter $\lambda = \sum_{k=1}^{n} (a_k - \bar{a})^2/\sigma^2$, where $\bar{a} = \sum_{k=1}^{n} a_k/n$.

Exercise 24.13 From (24.36), derive the mean and variance of the noncentral chi-square distribution with n degrees of freedom and noncentrality parameter $\lambda = \sum_{k=1}^{n} a_k^2$.

In a similar manner, we can define noncentral t and noncentral F distributions which are useful in studying the power properties of t and F tests. For example, in Eq. (24.32), we defined the F distribution with (m, n) degrees of freedom as the distribution of the variable

$$X = \frac{V/m}{W/n} = \frac{n}{m} \cdot \frac{V}{W} ,$$

where V and W are independent chi-square random variables with m and n degrees of freedom, respectively. Now, consider the distribution of the variable

$$X' = \frac{V'/m}{W'/n} = \frac{n}{m} \cdot \frac{V'}{W'} , \qquad (24.37)$$

where V' and W' are independent noncentral chi-square random variables with m and n degrees of freedom and noncentrality parameters λ_1 and λ_2, respectively. The distribution of X' is called the *doubly noncentral F distribution* with (m, n) degrees of freedom and noncentrality parameters (λ_1, λ_2). In the special case when $\lambda_2 = 0$ (i.e., when there is a central chi-square in the denominator), the distribution of X' is called the (singly) *noncentral F distribution* with (m, n) degrees of freedom and noncentrality parameter λ_1.

Part III

MULTIVARIATE DISTRIBUTIONS

CHAPTER 25

MULTINOMIAL DISTRIBUTION

25.1 Introduction

The multinomial distribution, being the multivariate generalization of the binomial distribution discussed in Chapter 5, is one of the most important and interesting multivariate discrete distributions.

Consider a sequence of independent and idential trials each of which can result in one of k possible mutually exclusive and collectively exhaustive events, say, $A_1, A_2, \ldots, A_k$, with respectively probabilities $p_1, p_2, \ldots, p_k$, where $p_1 + p_2 + \cdots + p_k = 1$. Such trials are termed *multinomial trials*. Let $\mathbf{Y}_\ell = (Y_{1,\ell}, Y_{2,\ell}, \ldots, Y_{k,\ell})$, $\ell = 1, 2, \ldots$, be the indicator vector variables, that is, $Y_{j,\ell}$ takes on the value 1 if the event A_j $(j = 1, 2, \ldots, k)$ is the outcome of the ℓth trial and the value 0 if the event A_j is not the outcome of the ℓth trial. Note that the variables $Y_{1,\ell}, Y_{2,\ell}, \ldots, Y_{k,\ell}$ (which are the components of the vector $\mathbf{Y}_\ell$) are dependent, and that

$$Y_{1,\ell} + Y_{2,\ell} + \cdots + Y_{k,\ell} = 1, \quad \ell = 1, 2, \ldots. \tag{25.1}$$

For any $n = 1, 2, \ldots$, let us now define the random vector $\mathbf{X}_n = (X_{1,n}, X_{2,n}, \ldots, X_{k,n})$ as

$$\mathbf{X}_n = \mathbf{Y}_1 + \mathbf{Y}_2 + \cdots + \mathbf{Y}_n, \quad n = 1, 2, \ldots, \tag{25.2}$$

where $X_{j,n} = Y_{j,1} + Y_{j,2} + \cdots + Y_{j,n}$ $(j = 1, 2, \ldots, k)$ is the number of occurrences of event A_j in the n multinomial trials. In other words, the random vector $\mathbf{X}_n$ is simply a counter which gives us the number of occurrences of the events $A_1, A_2, \ldots, A_k$ in the n multinomial trials, and hence

$$X_{1,n} + X_{2,n} + \cdots + X_{k,n} = n, \quad n = 1, 2, \ldots. \tag{25.3}$$

Then, simple probability arguments readily yield

$$
\begin{aligned}
P_n(m_1, m_2, \ldots, m_k) &= \Pr\{X_{1,n} = m_1, X_{2,n} = m_2, \ldots, X_{k,n} = m_k\} \\
&= \frac{n!}{m_1!\, m_2! \cdots m_k!} \prod_{r=1}^{k} p_r^{m_r},
\end{aligned}
$$
$$m_r = 0, \ldots, n, \ m_1 + \cdots + m_k = n. \tag{25.4}$$

25.2 Notations

A random vector $\mathbf{X}_n = (X_{1,n}, X_{2,n}, \ldots, X_{k,n})$ having the joint probability mass function as in (25.4) is said to have the *multinomial $M(n, p_1, p_2, \ldots, p_k)$ distribution*. In the case when $k = 3$, the distribution is also referred to as the *trinomial distribution*.

Remark 25.1 The random vector $\mathbf{X}_n \sim M(n, p_1, p_2, \ldots, p_k)$ is actually $(k-1)$-dimensional since its components $X_{1,n}, X_{2,n}, \ldots, X_{k,n}$ satisfy the relationship

$$X_{1,n} + X_{2,n} + \cdots + X_{k,n} = n$$

and, consequently, one of the components (say, $X_{k,n}$) can be expressed as

$$X_{k,n} = n - X_{1,n} - X_{2,n} - \cdots - X_{k-1,n}.$$

Hence, the distribution of the random vector $\mathbf{X}_n = (X_{1,n}, X_{2,n}, \ldots, X_{k,n})$ is completely determined by the distribution of the $(k-1)$-dimensional random vector $(X_{1,n}, X_{2,n}, \ldots, X_{k-1,n})$. For example, when $k = 2$, the probabilities $P_n(m_1, m_2)$ in (25.4) simply become

$$
\begin{aligned}
P_n(m_1, m_2) &= \frac{n!}{m_1!\, m_2!} p_1^{m_1} p_2^{m_2} \quad \text{(with } m_1 + m_2 = n \text{ and } p_1 + p_2 = 1\text{)} \\
&= \binom{n}{m_1} p_1^{m_1} (1 - p_1)^{n - m_1}, \quad m_1 = 0, 1, \ldots, n, \qquad (25.5)
\end{aligned}
$$

which are the binomial probabilities.

25.3 Compositions

Due to the probability interpretation of multinomial distributions given above, it readily follows that if independent vectors $\mathbf{Y}_1, \mathbf{Y}_2, \ldots, \mathbf{Y}_n$ all have multinomial $M(1, p_1, p_2, \ldots, p_k)$ distribution, then the sum $\mathbf{X}_n = \mathbf{Y}_1 + \mathbf{Y}_2 + \cdots + \mathbf{Y}_n$ has the multinomial $M(n, p_1, p_2, \ldots, p_k)$ distribution. In addition, if $\mathbf{X} \sim M(n_1, p_1, p_2, \ldots, p_k)$ and $\mathbf{Y} \sim M(n_2, p_1, p_2, \ldots, p_k)$ are independent multinomial random vectors, then $\mathbf{X} + \mathbf{Y}$ is distributed as the sum $\mathbf{Y}_1 + \mathbf{Y}_2 + \cdots + \mathbf{Y}_{n_1+n_2}$ of i.i.d. multinomial $M(1, p_1, p_2, \ldots, p_k)$ random vectors, and hence, is distributed as multinomial $M(n_1 + n_2, p_1, p_2, \ldots, p_k)$.

25.4 Marginal Distributions

The fact that the multinomial distribution is the joint distribution of the number of occurrences of the events $A_1, A_2, \ldots, A_k$ in n multinomial trials enables us to derive easily any marginal distribution of interest. Suppose that we are interested in finding the marginal probabilities

$$\Pr\{X_{1,n} = m_1, X_{2,n} = m_2, \ldots, X_{j,n} = m_j\}$$

for $j < k$, when $\mathbf{X}_n \sim M(n, p_1, p_2, \ldots, p_k)$. We first note that

$$\Pr\{X_{1,n} = m_1, \ldots, X_{j,n} = m_j\}$$
$$= \Pr\{X_{1,n} = m_1, \ldots, X_{j,n} = m_j, V = m\}, \qquad (25.6)$$

where

$$m = m_{j+1} + \cdots + m_k = n - m_1 - \cdots - m_j \text{ and } V = X_{j+1,n} + \cdots + X_{k,n};$$

evidently, V denotes the number of occurrences of the event $A = A_{j+1} \cup A_{j+2} \cup \cdots \cup A_k$ in the n multinomial trials with the corresponding probability of occurrence being

$$\Pr\{A\} = p_{j+1} + \cdots + p_k = 1 - p_1 - \cdots - p_j = p \quad (\text{say}) .$$

Then, the random vector $(X_{1,n}, \ldots, X_{j,n}, V)$ clearly has the multinomial $M(j+1, p_1, \ldots, p_j, p)$ distribution; then, using (25.4) and (25.6), we have

$$\Pr\{X_{1,n} = m_1, \ldots, X_{j,n} = m_j\} \quad = \quad P_n(m_1, \ldots, m_j, m)$$
$$= \quad \frac{n!}{m!\, m_1! \cdots m_j!} p^m \prod_{r=1}^{j} p_r^{m_r},$$
$$m + \sum_{i=1}^{j} m_i = n \text{ and } p + \sum_{i=1}^{j} p_i = 1.$$
$$(25.7)$$

In particular, for $j = 1$, we simply obtain from (25.7) the marginal distribution of $X_{1,n}$ as

$$\Pr\{X_{1,n} = m_1\} = \frac{n!}{m_1!\,(n-m_1)!} p_1^{m_1} (1 - p_1)^{n-m_1}, \ m_1 = 0, \ldots, n, \quad (25.8)$$

which simply reveals that the marginal distribution of $X_{1,n}$ is binomial $B(n, p_1)$. Similarly, we have $X_{r,n} \sim B(n, p_r)$ for any $r = 1, 2, \ldots, k$.

25.5   Conditional Distributions

Let $\mathbf{X}_n \sim M(n, p_1, p_2, \ldots, p_k)$. Consider now the conditional distribution of $(X_{j+1,n}, \ldots, X_{k,n})$, given $(X_{1,n} = m_1, \ldots, X_{j,n} = m_j)$, defined by

$$\Pr\{X_{j+1,n} = m_{j+1}, \ldots, X_{k,n} = m_k \mid X_{1,n} = m_1, \ldots, X_{j,n} = m_j\}$$
$$= \frac{\Pr\{X_{1,n} = m_1, \ldots, X_{k,n} = m_k\}}{\Pr\{X_{1,n} = m_1, \ldots, X_{j,n} = m_j\}}. \qquad (25.9)$$

Substituting the expressions in (25.4) and (25.7) into (25.9), we obtain

$$\Pr\{X_{j+1,n} = m_{j+1}, \ldots, X_{k,n} = m_k \mid X_{1,n} = m_1, \ldots, X_{j,n} = m_j\}$$

$$= \frac{\dfrac{n!}{m_1! \cdots m_k!} \displaystyle\prod_{r=1}^{k} p_r^{m_r}}{\dfrac{n!}{m! m_1! \cdots m_j!} p^m \displaystyle\prod_{r=1}^{j} p_r^{m_r}}$$

$$= \frac{m!}{m_{j+1}! \cdots m_k! p^m} \prod_{r=j+1}^{k} p_r^{m_r}$$

$$= \frac{(n - m_1 - \cdots - m_j)!}{m_{j+1}! \cdots m_k!} \prod_{r=j+1}^{k} \left(\frac{p_r}{p}\right)^{m_r} \tag{25.10}$$

for $m_{j+1} + \cdots + m_k = n - (m_1 + \cdots + m_j)$, and 0 otherwise. From (25.10), we readily observe that the conditional distribution of $(X_{j+1,n}, \ldots, X_{k,n})$, given $X_{1,n} = m_1, \ldots, X_{j,n} = m_j$, is multinomial $M(n - m, q_{j+1}, \ldots, q_k)$, where $m = m_1 + \cdots + m_j$, $q_i = p_i/p$ (for $i = j + 1, \ldots, k$), and $p = p_{j+1} + \cdots + p_k$. Since the dependence on $m_1, m_2, \ldots, m_j$ in (25.10) is only through the sum $m_1 + m_2 + \cdots + m_j$, we readily note that

$$\Pr\{X_{j+1,n} = m_{j+1}, \ldots, X_{k,n} = m_k \mid X_{1,n} = m_1, \ldots, X_{j,n} = m_j\}$$

$$= \Pr\left\{X_{j+1,n} = m_{j+1}, \ldots, X_{k,n} = m_k \;\middle|\; \sum_{i=1}^{j} X_{i,n} = \sum_{i=1}^{j} m_i\right\}; \tag{25.11}$$

hence, the conditional distribution of $(X_{j+1,n}, \ldots, X_{k,n})$, given $X_{1,n} + \cdots + X_{j,n} = m$, is also the same multinomial $M(n - m, q_{j+1}, \ldots, q_k)$ distribution.

25.6 Moments

Let $\mathbf{X}_n = (X_{1,n}, \ldots, X_{k,n}) \sim M(n, p_1, \ldots, p_k)$. Then, since the marginal distribution of $X_{r,n}$ $(r = 1, 2, \ldots, k)$ is binomial $B(n, p_r)$, we readily have

$$EX_{r,n} = np_r = e_r \quad \text{and} \quad \text{Var } X_{r,n} = np_r(1 - p_r) = \sigma_r^2. \tag{25.12}$$

Next, in order to derive the correlation between the variables $X_{r,n}$ $(r = 1, \ldots, k)$, we shall first find the covariance using the formula

$$\begin{aligned}
\sigma_{rs} &= E\{(X_{r,n} - EX_{r,n})(X_{s,n} - EX_{s,n})\} \\
&= E(X_{r,n} X_{s,n}) - e_r e_s \\
&= E\{X_{s,n} E(X_{r,n}|X_{s,n})\} - e_r e_s, \tag{25.13}
\end{aligned}$$

where the last equality follows from the fact that

$$E\left\{E\left(X_{r,n}\,X_{s,n}|X_{s,n}\right)\right\}$$
$$= \sum_j \sum_i ji \, \Pr\left\{X_{r,n}=i, X_{s,n}=j\right\}$$
$$= \sum_j j \, \Pr\left\{X_{s,n}=j\right\} \sum_i i \, \Pr\left\{X_{r,n}=i|X_{s,n}=j\right\}$$
$$= \sum_j j \, \Pr\left\{X_{s,n}=j\right\} E\left(X_{r,n}|X_{s,n}=j\right)$$
$$= E\left\{X_{s,n}\,E\left(X_{r,n}|X_{s,n}\right)\right\}. \tag{25.14}$$

Now, we shall explain how we can find the regression $E\left(X_{r,n}|X_{s,n}\right)$ required in (25.13). For the sake of simplicity, let us consider the case when $r=2$ and $s=1$. Using the fact that the conditional distribution of the vector $(X_{2,n},\ldots,X_{k,n})$, given $X_{1,n}=m$, is multinomial $M(n-m,q_2,\ldots,q_k)$, where $q_i = p_i/(1-p_1)$ $(i=2,\ldots,k)$, we readily have the conditional distribution of $X_{2,n}$, given $X_{1,n}=m$, is binomial $B(n-m,q_2)$; hence,

$$E\left(X_{2,n}|X_{1,n}=m\right) = (n-m)q_2 = \frac{(n-m)p_2}{1-p_1} \tag{25.15}$$

from which we obtain the regression of $X_{2,n}$ on $X_{1,n}$ to be $\dfrac{(n-X_{1,n})p_2}{1-p_1}$. Using this fact in (25.13), we obtain

$$
\begin{aligned}
\sigma_{12} &= E\left(X_{1,n}\,X_{2,n}\right) - e_1 e_2 \\
&= E\left\{X_{1,n}\,E\left(X_{2,n}|\xi_{1,n}\right)\right\} - e_1 e_2 \\
&= \frac{p_2}{1-p_1}E\left\{X_{1,n}\left(n-X_{1,n}\right)\right\} - e_1 e_2 \\
&= \frac{p_2}{1-p_1}\left\{n^2 p_1 - n p_1(1-p_1) - n^2 p_1^2\right\} - n^2 p_1 p_2 \\
&= -n p_1 p_2. \tag{25.16}
\end{aligned}
$$

Similarly, we have

$$\sigma_{rs} = \sigma_{sr} = -n p_r p_s \qquad \text{for any } 1 \le r < s \le k. \tag{25.17}$$

From (25.12) and (25.17), we thus have the covariance matrix of $\mathbf{X}_n = (X_{1,n},\ldots,$
$X_{k,n})$ to be

$$\sigma_{rs} = \begin{cases} n p_r(1-p_r) & \text{for } r = s \\ -n p_r p_s & \text{for } r \ne s \end{cases} \tag{25.18}$$

for $1 \le r, s \le k$. Furthermore, from (25.12) and (25.17), we also find the correlation coefficient between $X_{r,n}$ and $X_{s,n}$ $(1 \le r < s \le k)$ to be

$$\rho_{rs} = \frac{\sigma_{rs}}{\sigma_r \sigma_s} = -\sqrt{\frac{p_r p_s}{(1-p_r)(1-p_s)}}. \tag{25.19}$$

It is important to note that all the correlation coefficients are negative in a multinomial distribution.

Exercise 25.1 Derive the multiple regression function of $X_{r+1,n}$ on $X_{1,n}, \ldots, X_{r,n}$.

25.7 Generating Function and Characteristic Function

Consider a random vector $\mathbf{Y} = (Y_1, \ldots, Y_k)$ having $M(1, p_1, \ldots, p_k)$ distribution. As pointed out earlier, the distribution of this vector is determined by the nonzero probabilities (for $r = 1, \ldots, k$)

$$p_r = \Pr\{Y_1 = 0, \ldots, Y_{r-1} = 0, Y_r = 1, Y_{r+1} = 0, \ldots, Y_k = 0\}. \quad (25.20)$$

Then, it is evident that the generating function $Q(s_1, s_2, \ldots, s_k)$ of $\mathbf{Y}$ is

$$Q(s_1, s_2, \ldots, s_k) = E\left(s_1^{Y_1} \cdots s_k^{Y_k}\right) = \sum_{r=1}^{k} p_r s_r. \quad (25.21)$$

Since $\mathbf{X}_n \sim M(n, p_1, \ldots, p_k)$ is distributed as the sum $\mathbf{Y}_1 + \cdots + \mathbf{Y}_n$ [see Eq. (25.2)], where $\mathbf{Y}_1, \ldots, \mathbf{Y}_n$ are i.i.d. multinomial $M(1, p_1, \ldots, p_k)$ variables with generating function as in (25.21), we readily obtain the generating function of $\mathbf{X}_n$ as

$$
\begin{aligned}
P_n(s_1, \ldots, s_k) &= E\left(s_1^{X_{1,n}} \cdots s_k^{X_{k,n}}\right) \\
&= \{Q(s_1, \ldots, s_k)\}^n = \left(\sum_{r=1}^{k} p_r s_r\right)^n. \quad (25.22)
\end{aligned}
$$

From (25.22), we deduce the generating function of $(X_{1,n}, \ldots, X_{m,n})$ (for $m = 1, \ldots, k-1$) as

$$
\begin{aligned}
R_n(s_1, \ldots, s_m) &= P_n(s_1, \ldots, s_m, 1, \ldots, 1) \\
&= \left(\sum_{r=1}^{m} p_r s_r + \sum_{r=m+1}^{k} p_r\right)^n \\
&= \left\{1 + \sum_{r=1}^{m}(s_r - 1)p_r\right\}^n. \quad (25.23)
\end{aligned}
$$

In particular, when $m = 1$, we obtain from (25.23)

$$R_n(s_1) = E s_1^{X_{1,n}} = \{1 + (s_1 - 1)p_1\}^n, \quad (25.24)$$

which readily reveals that $X_{1,n}$ is distributed as binomial $B(n, p_1)$ (as noted earlier). Further, we obtain from (25.23) the generating function of the sum $X_{1,n} + \cdots + X_{m,n}$ (for $m = 1, \ldots, k-1$) as

$$Es^{X_{1,n} + \cdots + X_{m,n}} = R_n(s, \ldots, s) = \left\{ 1 + (s-1) \sum_{r=1}^{m} p_r \right\}^n, \qquad (25.25)$$

which reveals that the sum $X_{1,n} + \cdots + X_{m,n}$ is distributed as binomial $B\left(n, \sum_{r=1}^{m} p_r\right)$. Note that when $m = k$, the sum $X_{1,n} + \cdots + X_{k,n}$ has a degenerate distribution since $X_{1,n} + \cdots + X_{k,n} = n$.

Exercise 25.2 From the generating function of $\mathbf{X}_n \sim M(n, p_1, \ldots, p_k)$ in (25.22), establish the expressions of means, variances, and covariances derived in (25.12) and (25.17).

Exercise 25.3 From the generating function of $(X_{1,n}, \ldots, X_{m,n})$ in (25.23), prove that if $m > n$ and $m \le k$, then $E(X_{1,n} \cdots X_{m,n}) = 0$. Also, argue in this case that this expression must be true due to the fact that at least one of the $X_{r,n}$'s must be 0 since $X_{1,n} + \cdots + X_{k,n} = n$.

From (25.22), we immediately obtain the characteristic function of $\mathbf{X}_n \sim M(n, p_1, \ldots, p_k)$ as

$$
\begin{aligned}
f_n(t_1, \ldots, t_k) &= E\left\{ e^{i(t_1 X_{1,n} + \cdots + t_k X_{k,n})} \right\} \\
&= P_n\left(e^{it_1}, \ldots, e^{it_k} \right) \\
&= \left(i \sum_{r=1}^{k} p_r e^{it_r} \right)^n. \qquad (25.26)
\end{aligned}
$$

In addition, from (25.23), we readily obtain the characteristic function of $(X_{1,n}, \ldots, X_{m,n})$ (for $m = 1, \ldots, k-1$) as

$$
\begin{aligned}
g_n(t_1, \ldots, t_m) &= R_n\left(e^{it_1}, \ldots, e^{it_m} \right) \\
&= \left\{ 1 + \sum_{r=1}^{m} p_r \left(e^{it_r} - 1 \right) \right\}^n. \qquad (25.27)
\end{aligned}
$$

Exercise 25.4 From (25.27), deduce the characteristic function of the sum $X_{1,n} + \cdots + X_{m,n}$ (for $m = 1, 2, \ldots, k-1$) and show that it corresponds to that of the binomial $B\left(n, \sum_{r=1}^{m} p_r\right)$.

25.8　Limit Theorems

Let us now consider the sequence of random vectors

$$\mathbf{X}_n = (X_{1,n}, \ldots, X_{k,n}) \sim M(n, p_1, \ldots, p_k), \tag{25.28}$$

where $p_k = 1 - \sum_{r=1}^{k-1} p_r$. Let $p_r = \lambda_r/n$ for $r = 1, \ldots, k-1$. Then, for $m = k-1$, the characteristic function of $(X_{1,n}, \ldots, X_{k-1,n})$ in (25.27) becomes

$$g_n(t_1, \ldots, t_{k-1}) = \left\{ 1 + \frac{1}{n} \sum_{r=1}^{k-1} \lambda_r \left(e^{it_r} - 1 \right) \right\}^n. \tag{25.29}$$

Lettting $n \to \infty$ in (25.29), we observe that

$$g_n(t_1, \ldots, t_{k-1}) \to \exp \left\{ \sum_{r=1}^{k-1} \lambda_r \left(e^{it_r} - 1 \right) \right\} = \prod_{r=1}^{k-1} h_r(t_r), \tag{25.30}$$

where $h_r(t) = \exp \left\{ \lambda_r \left(e^{it} - 1 \right) \right\}$ is the characteristic function of the Poisson $\pi(\lambda_r)$ distribution (for $r = 1, \ldots, k-1$). Hence, we observe from (25.30) that, as $n \to \infty$, the components $X_{1,n}, \ldots, X_{k-1,n}$ of the multinomial random vector $\mathbf{X}_n$ in (25.28) are asymptotically independent and that the marginal distribution of $X_{r,n}$ converges to the Poisson $\pi(\lambda_r)$ distribution for any $r = 1, 2, \ldots, k-1$.

Exercise 25.5 Using a similar argument, show that $X_{1,n} + \cdots + X_{m,n}$ (for $m = 1, \ldots, k-1$) converges to the Poisson $\pi \left(\sum_{r=1}^{m} \lambda_r \right)$ distribution.

Next, let us consider the sequence of the $(k-1)$-dimensional random vectors

$$\mathbf{W}_n = \left(\frac{X_{1,n} - np_1}{\sqrt{np_1(1-p_1)}}, \ldots, \frac{X_{k-1,n} - np_{k-1}}{\sqrt{np_{k-1}(1-p_{k-1})}} \right), n = 1, 2, \ldots. \tag{25.31}$$

Let $h_n(t_1, \ldots, t_{k-1})$ be the characteristic function of $\mathbf{W}_n$ in (25.31). Then, it follows from (25.27) that

$$h_n(t_1, \ldots, t_{k-1}) = \exp \left\{ -i\sqrt{n} \sum_{r=1}^{k-1} \frac{\sqrt{p_r}\,t_r}{\sqrt{1-p_r}} \right\}$$

$$\times \left[1 + \sum_{r=1}^{k-1} p_r \left\{ \exp \left(\frac{t_r}{\sqrt{np_r(1-p_r)}} \right) - 1 \right\} \right]^n. \tag{25.32}$$

Exercise 25.6 As $n \to \infty$, show that $h_n(t_1, \ldots, t_{k-1})$ in (25.32) converges to

$$h(t_1, \ldots, t_{k-1}) = \exp\left\{-\frac{1}{2}\left(\sum_{r=1}^{k-1} t_r^2 + 2 \sum_{1 \leq r < s \leq k-1} \rho_{rs} t_r t_s\right)\right\}, \quad (25.33)$$

where

$$\rho_{rs} = -\sqrt{\frac{p_r p_s}{(1 - p_r)(1 - p_s)}}$$

is the correlation coefficient between $X_{r,n}$ and $X_{s,n}$ derived in (25.19).

From (25.33), we see that the limiting characteristic function of the random variable

$$W_{r,n} = \frac{X_{r,n} - n p_r}{\sqrt{n p_r (1 - p_r)}}$$

becomes $\exp\left(-\frac{1}{2} t_r^2\right)$, which readily implies that the limiting distribution of the random variable $W_{r,n}$ is indeed standard normal (for $r = 1, \ldots, k-1$).

Furthermore, in Chapter 26, we will see that the limiting characteristic function of $\mathbf{W}_n$ in (25.33) corresponds to that of a multivariate normal distribution with mean vector $(0, \ldots, 0)$ and covariance matrix

$$\sigma_{ij} = \begin{cases} 1 & \text{if } i = j \\ -\sqrt{\dfrac{p_i p_j}{(1 - p_i)(1 - p_j)}} & \text{if } i \neq j \end{cases} \quad (25.34)$$

for $1 \leq i, j \leq k-1$. Hence, we have the asymptotic distribution of the random vector $\mathbf{W}_n$ in (25.31) to be multivariate normal.

CHAPTER 26

MULTIVARIATE NORMAL DISTRIBUTION

26.1 Introduction

The multivariate normal distribution is the most important and interesting multivariate distribution and based on it, a huge body of multivariate analysis has been developed. In this chapter we present a brief description of the multivariate normal distribution and some of its basic properties. For a detailed discussion on multivariate normal distribution and its properties, one may refer to the book by Tong (1990).

At the end of Chapter 25 (see Exercise 25.6), we found that the limiting distribution of a sequence of multinomial random variables has its characteristic function as [see Eq. (25.33)]

$$h(t_1,\ldots,t_{k-1}) = \exp\left\{-\frac{1}{2}\,Q\,(t_1,\ldots,t_{k-1})\right\}, \tag{26.1}$$

where $Q\,(t_1,\ldots,t_{k-1})$ is the quadratic form

$$Q\,(t_1,\ldots,t_{k-1}) = \sum_{r=1}^{k-1} t_r^2 + 2 \sum_{1 \le r < s \le k-1} \rho_{rs} t_r t_s. \tag{26.2}$$

The quadratic form $Q\,(t_1,\ldots,t_{k-1})$ in (26.2) can be written in matrix notation as

$$Q(\mathbf{t}) = \mathbf{t}\Sigma\mathbf{t}', \tag{26.3}$$

where $\mathbf{t}$ is a row vector $(t_1,\ldots,t_{k-1})$, and Σ is a $(k-1)\times(k-1)$ real symmetric positive definite matrix with (i,i)th element as 1 and (i,j)th element as ρ_{ij}. In fact, Σ is the covariance matrix of a random vector $(Y_1,\ldots,Y_{k-1})$ with variances 1 and covariances ρ_{ij} (which are also the correlation coefficients).

It can be shown that (we state this result without proof)

$$\exp\left\{-\frac{1}{2}\,Q\,(t_1,\ldots,t_{k-1})\right\}$$

259

$$= \int_{-\infty}^{\infty} \cdots \int_{-\infty}^{\infty} e^{i(t_1 x_1 + \cdots + t_{k-1} x_{k-1})} \, p(x_1, \ldots, x_{k-1}) \, dx_1 \cdots dx_{k-1},$$

$$(26.4)$$

where

$$p(x_1, \ldots, x_{k-1}) = \frac{1}{\sqrt{(2\pi)^{k-1}|\Sigma|}} \exp\left\{ -\frac{1}{2} \, Q^{-1}(x_1, \ldots, x_{k-1}) \right\} \quad (26.5)$$

with $|\Sigma|$ denoting the determinant of the matrix Σ, $Q^{-1}(\mathbf{t}) = \mathbf{t}\Sigma^{-1}\mathbf{t}'$, and Σ^{-1} denoting the inverse of the matrix Σ.

Equation (26.4) implies that the nonnegative function $p(x_1, \ldots, x_{k-1})$ satisfies the condition

$$\int_{-\infty}^{\infty} \cdots \int_{-\infty}^{\infty} p(x_1, \ldots, x_{k-1}) \, dx_1 \cdots dx_{k-1} \;=\; \exp\left\{ -\frac{1}{2} \, Q(0, \ldots, 0) \right\}$$

$$= \; 1 \qquad\qquad (26.6)$$

using (26.2). Hence, $p(x_1, \ldots, x_{k-1})$ is the pdf of some $(k-1)$-dimensional random variable, and

$$h(t_1, \ldots, t_{k-1}) = \exp\left\{ -\frac{1}{2} \, Q(t_1, \ldots, t_{k-1}) \right\} \qquad (26.7)$$

is indeed the characteristic function of this distribution.

26.2   Notations

Let $\Sigma = (\sigma_{ij})_{i,j=1}^{n}$ be a $(n \times n)$ real symmetric positive definite matrix. Note that such a matrix may be a matrix of second-order moments of some n-dimensional distribution. Let us now define the quadratic form corresponding to Σ as

$$Q(\mathbf{t}) = \mathbf{t}\Sigma\mathbf{t}' = \sum_{r=1}^{n} \sigma_{rr} t_r^2 + 2 \sum_{1 \le r < s \le n} \sigma_{rs} t_r t_s. \qquad (26.8)$$

As before, let Σ^{-1} denote the inverse of the matrix Σ and $Q^{-1}(\mathbf{t}) = \mathbf{t}\Sigma^{-1}\mathbf{t}'$.

Then, a random vector $\mathbf{Y} = (Y_1, \ldots, Y_n)$ is said to have a *multivariate normal* $MN(\mathbf{0}, \Sigma)$ *distribution* if its pdf is of the form

$$g(x_1, \ldots, x_n) = \frac{1}{\sqrt{(2\pi)^n|\Sigma|}} \exp\left\{ -\frac{1}{2} \, Q^{-1}(x_1, \ldots, x_n) \right\}, \qquad (26.9)$$

and its characteristic function is

$$h(t_1, \ldots, t_n) = \exp\left\{ -\frac{1}{2} \, Q(t_1, \ldots, t_n) \right\}, \qquad (26.10)$$

where the quadratic form $Q(t_1, \ldots, t_n)$ is as in (26.8). In this case, the first parameter $\mathbf{0}$ in the notation $MN(\mathbf{0}, \Sigma)$ is a row vector of dimension $(n \times 1)$

with all of its elements being 0. From the characteristic function of $\mathbf{Y}$ in (26.10), we then readily find that

$$EY_r = 0 \quad \text{for } r = 1, \ldots, n, \qquad (26.11)$$

and that Σ is indeed the matrix of the second-order moments of $\mathbf{Y}$, that is,

$$\begin{aligned}
\text{Var } Y_r &= \sigma_{rr} \quad \text{for } 1 \leq r \leq n, \\
\text{Cov}(Y_r, Y_s) &= \sigma_{rs} \quad \text{for } 1 \leq r < s \leq n.
\end{aligned} \qquad (26.12)$$

Using the random vector $\mathbf{Y}$, we can now introduce a new random vector

$$\mathbf{X} = (X_1, \ldots, X_n) = (Y_1 + m_1, \ldots, Y_n + m_n) = \mathbf{Y} + \mathbf{m}, \qquad (26.13)$$

where $\mathbf{m} = (m_1, \ldots, m_n)$ is the vector of means of the components of the random vector $\mathbf{X}$.

Exercise 26.1 From (26.9), show that the pdf of the random vector $\mathbf{X}$ is given by

$$p(x_1, \ldots, x_n) = \frac{1}{\sqrt{(2\pi)^n |\Sigma|}} \exp\left\{ -\frac{1}{2} Q^{-1}(x_1 - m_1, \ldots, x_n - m_n) \right\}. \tag{26.14}$$

Exercise 26.2 From (26.10), show that the characteristic function of the random vector $\mathbf{X}$ is given by

$$\begin{aligned}
f(t_1, \ldots, t_n) &= E e^{it\mathbf{X}'} \\
&= \exp\left\{ i \sum_{r=1}^{n} m_r t_r - \frac{1}{2} \left(\sum_{r=1}^{n} \sigma_{rr} t_r^2 + 2 \sum_{1 \leq r < s \leq n} \sigma_{rs} t_r t_s \right) \right\} \\
&= \exp\left\{ it\mathbf{m}' - \frac{1}{2} Q(t_1, \ldots, t_n) \right\}, \qquad (26.15)
\end{aligned}$$

where the quadratic form $Q(t_1, \ldots, t_n)$ is as in (26.8).

We then say that such a random vector $\mathbf{X} = (X_1, \ldots, X_n)$ has a *multivariate normal distribution* (in n dimensions, of course) with mean vector $\mathbf{m}$ and covariance matrix Σ, and we denote it by $\mathbf{X} \sim MN(\mathbf{m}, \Sigma)$.

26.3   Marginal Distributions

Let $\mathbf{X} = (X_1, \ldots, X_n) \sim MN(\mathbf{m}, \Sigma)$, and $\mathbf{V} = (X_1, \ldots, X_\ell)$, $\ell < n$. Then, from the characteristic function of $\mathbf{X}$ in (26.15), we readily obtain the characteristic function of $\mathbf{V}$ as

$$Ee^{i(t_1 X_1 + \cdots + t_\ell X_\ell)}$$

$$= f(t_1, \ldots, t_r, 0, \ldots, 0)$$

$$= \exp\left\{ i \sum_{r=1}^{\ell} m_r t_r - \frac{1}{2} \left(\sum_{r=1}^{\ell} \sigma_{rr} t_r^2 + 2 \sum_{1 \le r < s \le \ell} \sigma_{rs} t_r t_s \right) \right\}.$$

$$(26.16)$$

Equation (26.16) immediately implies that the random vector $\mathbf{V} = (X_1, \ldots, X_\ell)$, $\ell < n$, is distributed as $MN(\mathbf{m}^{(\ell)}, \Sigma^{(\ell)})$, where

$$\mathbf{m}^{(\ell)} = (m_1, \ldots, m_\ell) \quad \text{and} \quad \Sigma^{(\ell)} = (\sigma_{ij})_{i,j=1}^{\ell}.$$

In particular, we observe that marginally $X_1 \sim N(m_1, \sigma_{11})$, where $m_1 = EX_1$ and $\sigma_{11} = \text{Var } X_1$. Similarly, it can be shown that marginally, $X_r \sim N(m_r, \sigma_{rr})$ for any $r = 1, \ldots, n$.

26.4   Distributions of Sums

Let $\mathbf{X}$ and $\mathbf{Y}$ be two independent (n-dimensional) multivariate normal random vectors with parameters $(\mathbf{m}^{(1)}, \Sigma^{(1)})$ and $(\mathbf{m}^{(2)}, \Sigma^{(2)})$, respectively. Further, let $\mathbf{V}$ be a new random vector defined as $\mathbf{V} = \mathbf{X} + \mathbf{Y}$.

Exercise 26.3 Then, using Eq. (26.15), prove that $\mathbf{V} \sim MN(\mathbf{m}^{(1)} + \mathbf{m}^{(2)}, \Sigma^{(1)} + \Sigma^{(2)})$. More generally, if $\mathbf{X}_j \sim MN(\mathbf{m}^{(j)}, \Sigma^{(j)})$ $(j = 1, \ldots, k)$ are independent random vectors, prove that $\sum_{j=1}^{k} \mathbf{X}_j \sim MN(\mathbf{m}, \Sigma)$, where

$$\mathbf{m} = \sum_{j=1}^{k} \mathbf{m}^{(j)} \quad \text{and} \quad \Sigma = \sum_{j=1}^{k} \Sigma^{(j)}.$$

26.5   Linear Combinations of Components

Let $\mathbf{X} = (X_1, \ldots, X_n) \sim MN(\mathbf{m}, \Sigma)$. Now, let us consider the linear combination $L = \sum_{r=1}^{n} c_r X_r = \mathbf{X}\mathbf{c}'$, where $\mathbf{c} = (c_1, \ldots, c_n)$. Then, from the characteristic function of $\mathbf{X}$ in (26.15), we readily obtain the characteristic function of L as

$$
\begin{aligned}
f_L(t) &= Ee^{itL} \\
&= E\left[\exp\left\{it\sum_{r=1}^{n}c_rX_r\right\}\right] \\
&= f(c_1t,\ldots,c_nt) \\
&= \exp\left\{it\sum_{r=1}^{n}m_rc_r - \frac{1}{2}t^2\left(\sum_{r=1}^{n}\sigma_{rr}c_r^2 + 2\sum_{1\leq r<s\leq n}\sigma_{rs}c_rc_s\right)\right\}.
\end{aligned}
$$

(26.17)

Equation (26.17) readily reveals that the linear combination L is distributed as normal $N(m,\sigma^2)$, where

$$
m = \sum_{r=1}^{n}c_rm_r = \mathbf{mc}'
$$

and

$$
\begin{aligned}
\sigma^2 &= \sum_{r=1}^{n}\sigma_{rr}c_r^2 + 2\sum_{1\leq r<s\leq n}\sigma_{rs}c_rc_s \\
&= \sum_{r=1}^{n}c_r^2\,\mathrm{Var}\,\xi_r + 2\sum_{1\leq r<s\leq n}c_rc_s\,\mathrm{Cov}(\xi_r,\xi_s) \\
&= \mathbf{c}\Sigma\mathbf{c}'\,.
\end{aligned}
$$

26.6   Independence of Components

It is easy to find conditions under which the components of the multivariate normal vector $\mathbf{X} = (X_1,\ldots,X_n) \sim MN(\mathbf{m},\Sigma)$ are all independent. Since in this case, when $X_1,\ldots,X_n$ are all independent, the characteristic function of $\mathbf{X}$ in (26.15) must satisfy the condition

$$
f(t_1,\ldots,t_n) = \prod_{r=1}^{n}f(0,\ldots,0,t_r,0,\ldots,0),
$$

(26.18)

we immediately observe that the components of the random vector $\mathbf{X}$ are independent if and only if

$$
\sigma_{ij} = 0 \quad \text{for } 1 \leq i < j \leq n.
$$

(26.19)

In other words, the components $X_1,\ldots,X_n$ are all independent if and only if the covariance matrix Σ is a diagonal matrix.

26.7　Linear Transformations

Let $\mathbf{X} = (X_1, \ldots, X_n) \sim MN(\mathbf{m}, \Sigma)$. Further, let $\mathbf{Y} = (Y_1, \ldots, Y_n) = \mathbf{XC}$ be a linear transformation of $\mathbf{X}$, where $\mathbf{C}$ is a $(n \times n)$ non-singular matrix.

Exercise 26.4 Then, prove that $\mathbf{Y} = (Y_1, \ldots, Y_n) \sim MN(\mathbf{mC}, \mathbf{C}'\Sigma\mathbf{C})$.

It is well known from matrix theory that for any symmetric $(n \times n)$ matrix Σ, there exists an $(n \times n)$ orthogonal matrix $\mathbf{C}$ such that $\mathbf{C}'\Sigma\mathbf{C}$ is a diagonal matrix; be reminded that a matrix $\mathbf{C}$ is said to be an orthogonal matrix if $\mathbf{CC}' = \mathbf{I}_n$, where $\mathbf{I}_n$ denotes an identity matrix of order $(n \times n)$. From this property, it is clear that if $\mathbf{C}'\Sigma\mathbf{C}$ is a diagonal matrix, then the components of the random vector $\mathbf{Y} = (Y_1, \ldots, Y_n) \sim MN(\mathbf{mC}, \mathbf{C}'\Sigma\mathbf{C})$ will all be independent. Hence, if $\mathbf{X} = (X_1, \ldots, X_n) \sim MN(\mathbf{m}, \Sigma)$, then there exists an orthogonal linear transformation $\mathbf{Y} = \mathbf{XC}$ that generates independent normal random variables

$$Y_r = c_{1r}X_1 + \cdots + c_{rn}X_n, \qquad r = 1, \ldots, n. \tag{26.20}$$

Moreover, we see that $\mathbf{X}$ can be expressed as $\mathbf{X} = \mathbf{YB}$, where $\mathbf{B} = \mathbf{C}^{-1}$ is also an orthogonal matrix. This implies that the components $X_1, \ldots, X_n$ of any multivariate normal random vector $\mathbf{X}$ can be expressed as linear combinations

$$X_r = b_{r1}Y_1 + \cdots + b_{rn}Y_n, \qquad r = 1, \ldots, n \tag{26.21}$$

of independent normal random variables $Y_1, \ldots, Y_n$.

The orthogonality of the transformations in (26.20) and (26.21) can be exploited to obtain the following relations:

$$\sum_{r=1}^{n} Y_r^2 = \mathbf{YY}' = \mathbf{XC}\,(\mathbf{XC})' = \mathbf{X}(\mathbf{CC}')\mathbf{X}' = \mathbf{XI}_n\mathbf{X}' = \mathbf{XX}' = \sum_{r=1}^{n} X_r^2 \tag{26.22}$$

and

$$\begin{aligned}
\sum_{r=1}^{n}(Y_r - EY_r)^2 &= (\mathbf{Y} - E\mathbf{Y})\,(\mathbf{Y} - E\mathbf{Y})' \\
&= (\mathbf{XC} - E\mathbf{XC})\,(\mathbf{XC} - E\mathbf{XC})' \\
&= (\mathbf{X} - E\mathbf{X})\,\mathbf{CC}'\,(\mathbf{X} - E\mathbf{X})' \\
&= (\mathbf{X} - E\mathbf{X})\,\mathbf{I}_n\,(\mathbf{X} - E\mathbf{X})' \\
&= (\mathbf{X} - E\mathbf{X})\,(\mathbf{X} - E\mathbf{X})' \\
&= \sum_{r=1}^{n}(X_r - EX_r)^2.
\end{aligned} \tag{26.23}$$

It follows, for example, from (26.22) and (26.23) that

$$\sum_{r=1}^{n} EY_r^2 = \sum_{r=1}^{n} EX_r^2 \qquad (26.24)$$

and

$$\sum_{r=1}^{n} \text{Var } Y_r = \sum_{r=1}^{n} \text{Var } X_r. \qquad (26.25)$$

26.8   Bivariate Normal Distribution

In this section we discuss in more detail the special case when $n = 2$, that is, the two-dimensional random vector (X_1, X_2) having a bivariate normal distribution. In this case, we denote the mean vector by (m_1, m_2) and the covariance matrix by

$$\Sigma = \begin{pmatrix} \sigma_1^2 & \rho\sigma_1\sigma_2 \\ \rho\sigma_1\sigma_2 & \sigma_2^2 \end{pmatrix}, \qquad (26.26)$$

where $\sigma_k^2 = \text{Var } X_k$ $(k = 1, 2)$, $\rho\sigma_1\sigma_2 = \text{Cov}(X_1, X_2)$, and ρ is the correlation coefficient between X_1 and X_2. Since this bivariate normal distribution depends on five parameters, we shall denote it by $BN(m_1, m_2, \sigma_1^2, \sigma_2^2, \rho)$.

Characteristic function:
    The characteristic function for this special case is deduced from (26.15) to be

$$h(t_1, t_2) = \exp\left\{ i\left(m_1 t_1 + m_2 t_2\right) - \frac{1}{2}\left(\sigma_1^2 t_1^2 + 2\rho\sigma_1\sigma_2 t_1 t_2 + \sigma_2^2 t_2^2\right) \right\}. \qquad (26.27)$$

Note that the expressions of the mean vector and the covariance matrix given above can also be obtained easily from the characteristic function in (26.27).

Density function:
    From (26.26), we note that the determinant of the matrix Σ is $|\Sigma| = \sigma_1^2\sigma_2^2(1 - \rho^2)$, which is positive if $\sigma_1^2 > 0$, $\sigma_2^2 > 0$, and $|\rho| < 1$. If $|\rho| = 1$, then there exists a linear dependence between the variables X_1 and X_2 and, therefore, the vector (X_1, X_2) has a degenerate normal distribution.
    The inverse of the matrix Σ in (26.26) can easily be shown to be

$$\Sigma^{-1} = \frac{1}{\sigma_1^2\sigma_2^2(1 - \rho^2)} \begin{pmatrix} \sigma_2^2 & -\rho\sigma_1\sigma_2 \\ -\rho\sigma_1\sigma_2 & \sigma_1^2 \end{pmatrix}. \qquad (26.28)$$

Upon substituting for these expressions in (26.14) and simplifying, we obtain the pdf of (X_1, X_2) to be

$$p(x_1, x_2) = \frac{1}{2\pi\sigma_1\sigma_2\sqrt{1 - \rho^2}} \exp\left[-\frac{1}{2(1 - \rho^2)} \left\{ \left(\frac{x_1 - m_1}{\sigma_1}\right)^2 \right.\right.$$

$$-2\rho\left(\frac{x_1-m_1}{\sigma_1}\right)\left(\frac{x_2-m_2}{\sigma_2}\right)+\left(\frac{x_2-m_2}{\sigma_2}\right)^2\bigg\}\bigg].$$

$$(26.29)$$

Exercise 26.5 By integrating the pdf $p(x_1,x_2)$ in (26.29) with respect to x_2 and x_1, show that the marginal distributions of X_1 and X_2 are $N(m_1,\sigma_1^2)$ and $N(m_2,\sigma_2^2)$, respectively.

It should, however, be mentioned here that if the marginal distributions of X_1 and X_2 are both normal, it does not necessarily imply that the vector (X_1,X_2) should be distributed as bivariate normal. In order to see this, let us consider the bivariate density function

$$g(x_1,x_2) \;=\; \frac{1}{4\pi\sqrt{1-\rho^2}}\left[\exp\left\{-\frac{1}{2(1-\rho^2)}\left(x_1^2+2\rho x_1 x_2+x_2^2\right)\right\}\right.$$
$$\left.+\exp\left\{-\frac{1}{2(1-\rho^2)}\left(x_1^2-2\rho x_1 x_2+x_2^2\right)\right\}\right],$$

$$(26.30)$$

which is a mixture of the densities of $BN(0,0,1,1,-\rho)$ and $BN(0,0,1,1,\rho)$.

Exercise 26.6 If the bivariate random vector (X_1,X_2) has its pdf as in (26.30), then show that $X_1 \sim N(0,1)$, $X_2 \sim N(0,1)$, and $\mathrm{Cov}(X_1,X_2)=0$.

But, as seen earlier, among all bivariate normal distributions, only $BN(0,0,1,1,0)$ with pdf

$$h(x_1,x_2) = \frac{1}{2\pi}\,\exp\left\{-\frac{1}{2}\left(x_1^2+x_2^2\right)\right\}\qquad(26.31)$$

can provide such properties for the marginal distributions. Since the pdf $h(x_1,x_2)$ in (26.31) does not equal the pdf $g(x_1,x_2)$ in (26.30), we can conclude that $g(x_1,x_2)$ in (26.30) is not a bivariate normal density function, but it does have both its marginal distributions to be normal.

Some relationships:
    Let V and W be independent standard normal variables. Further, let

$$X_1 = V \quad\text{and}\quad X_2 = \rho V + \sqrt{1-\rho^2}\,W,\qquad(26.32)$$

where $|\rho| < 1$. Then, we readily have the characteristic function of the bivariate random vector (X_1, X_2) to be

$$
\begin{aligned}
f(t_1, t_2) &= E e^{it_1 X_1 + it_2 X_2} \\
&= E \exp \left\{ i(t_1 + t_2 \rho) V + i t_2 \sqrt{1 - \rho^2} W \right\} \\
&= E e^{i(t_1 + t_2 \rho) V} \, E e^{i t_2 \sqrt{1 - \rho^2} W} \\
&= \exp \left\{ - \frac{1}{2}(t_1 + t_2 \rho)^2 - \frac{1}{2} t_2^2 (1 - \rho^2) \right\} \\
&= \exp \left\{ - \frac{1}{2} \left(t_1^2 + 2\rho t_1 t_2 + t_2^2 \right) \right\}, \tag{26.33}
\end{aligned}
$$

which, when compared with (26.27), readily implies that (X_1, X_2) is distributed as $BN(0, 0, 1, 1, \rho)$.

Exercise 26.7 Establish this result by using the Jacobian method on the density function.

Exercise 26.8 More generally, prove that the bivariate random vector (X_1, X_2), where

$$
X_1 = m_1 + \sigma_1 V \quad \text{and} \quad X_2 = m_2 + \sigma_2 \left(\rho V + \sqrt{1 - \rho^2} W \right), \tag{26.34}
$$

is distributed as $BN(m_1, m_2, \sigma_1^2, \sigma_2^2, \rho)$.

Conditional distributions:

Let the random vector (X_1, X_2) be distributed as $BN(m_1, m_2, \sigma_1^2, \sigma_2^2, \rho)$. In this case, as seen earlier, $X_1 \sim N(m_1, \sigma_1^2)$ and $X_2 \sim N(m_2, \sigma_2^2)$. From (26.29), we can then obtain the conditional density function of X_1, given $X_2 = x_2$, as

$$
\begin{aligned}
&p_{X_1 | X_2}(x_1 | x_2) \\
&= \frac{p(x_1, x_2)}{p_{X_2}(x_2)} \\
&= \frac{\sqrt{2\pi} \sigma_2}{2\pi \sigma_1 \sigma_2 \sqrt{1 - \rho^2}} \exp \left[- \frac{1}{2(1 - \rho^2)} \left\{ \left(\frac{x_1 - m_1}{\sigma_1} \right)^2 \right. \right. \\
&\qquad \left. - 2\rho \left(\frac{x_1 - m_1}{\sigma_1} \right) \left(\frac{x_2 - m_2}{\sigma_2} \right) + \left(\frac{x_2 - m_2}{\sigma_2} \right)^2 \right\} \\
&\qquad \left. + \frac{1}{2} \left(\frac{x_2 - m_2}{\sigma_2} \right)^2 \right]
\end{aligned}
$$

$$= \frac{1}{\sqrt{2\pi}\sigma_1\sqrt{1-\rho^2}} \exp\left[-\frac{1}{2(1-\rho^2)}\left\{\left(\frac{x_1-m_1}{\sigma_1}\right)^2\right.\right.$$

$$\left.\left.- 2\rho\left(\frac{x_1-m_1}{\sigma_1}\right)\left(\frac{x_2-m_2}{\sigma_2}\right) + \left(\frac{\rho(x_2-m_2)}{\sigma_2}\right)^2\right\}\right]$$

$$= \frac{1}{\sqrt{2\pi}\sigma_1\sqrt{1-\rho^2}} \exp\left\{-\frac{1}{2\sigma_1^2(1-\rho^2)}(x_1-\lambda(x_2))^2\right\},$$

$$(26.35)$$

where

$$\lambda(x_2) = m_1 + \rho\sigma_1\left(\frac{x_2-m_2}{\sigma_2}\right). \qquad (26.36)$$

From (26.35), it is clear that the conditional distribution of X_1, given $X_2 = x_2$, is simply $N(\lambda(x_2), \sigma_1^2(1-\rho^2))$, where $\lambda(x_2)$ is as given in (26.36).

Exercise 26.9 Proceeding similarly, prove that the conditional distribution of X_2, given $X_1 = x_1$, is $N(\lambda^*(x_1), \sigma_2^2(1-\rho^2))$, where

$$\lambda^*(x_1) = m_2 + \rho\sigma_2\left(\frac{x_1-m_1}{\sigma_1}\right). \qquad (26.37)$$

Regressions:
   From Eq. (26.35), we immediately have the conditional mean of X_1 to be

$$E\left(X_1|X_2 = x_2\right) = \lambda(x_2) = m_1 + \rho\sigma_1\left(\frac{x_2-m_2}{\sigma_2}\right), \qquad (26.38)$$

which is a linear function in x_2. Similarly, we have from Exercise 26.9 that

$$E\left(X_2|X_1 = x_1\right) = \lambda^*(x_1) = m_2 + \rho\sigma_2\left(\frac{x_1-m_1}{\sigma_1}\right), \qquad (26.39)$$

which is a linear function in x_1. Thus, for a bivariate normal distribution, both regression functions (of X_1 on X_2 and of X_2 on X_1) are linear.
   Furthermore, we note that the conditional variances are at most as large as the corresponding unconditional variances. In fact, the conditional variances are strictly smaller than the corresponding unconditional variances whenever the correlation coefficient $|\rho| > 0$.

CHAPTER 27

DIRICHLET DISTRIBUTION

27.1 Introduction

In Chapter 18, when dealing with exponential order statistics, we established an important property of order statistics $U_{1,n} \leq U_{2,n} \leq \cdots \leq U_{n,n}$ as [see Eq. (18.25)]

$$\{U_{k,n}\}_{k=1}^{n} \stackrel{d}{=} \left\{ \frac{S_k}{S_{n+1}} \right\}_{k=1}^{n}, \tag{27.1}$$

where $S_k = Y_1 + Y_2 + \cdots + Y_k$ $(k = 1, 2, \ldots, n+1)$ is a sum of independent and identically distributed standard exponential random variables. Now, let $T_{k,n} = U_{k,n} - U_{k-1,n}$ $(k = 1, 2, \ldots, n)$, with the convention that $U_{0,n} \equiv 0$, denote the uniform spacings. Then, from (27.1) it is evident that

$$\{T_{k,n}\}_{k=1}^{n} \stackrel{d}{=} \left\{ \frac{Y_k}{S_{n+1}} \right\}_{k=1}^{n}. \tag{27.2}$$

Clearly, therefore, the joint distribution of the uniform spacings $T_{1,n}, \ldots, T_{n,n}$ is the same as the joint distribution of $Y_1/S_{n+1}, \ldots, Y_n/S_{n+1}$. In the simplest case of $k = n = 1$, it is clear that we have only one variable $Y_1/(Y_1 + Y_2)$ and that its distribution is the same as that of $T_{1,1} = U_{1,1} = U_1$; in other words, $Y_1/(Y_1 + Y_2)$ has the standard uniform $U(0, 1)$ distribution. In the general case, however, the distribution is somewhat involved and we will now proceed to derive it.

With $Y_1, Y_2, \ldots, Y_{n+1}$ being independent standard exponential random variables, we have their joint density function as

$$p_{Y_1,\ldots,Y_{n+1}}(y_1, \ldots, y_{n+1}) = e^{-(y_1 + \cdots + y_{n+1})}, \quad y_1, \ldots, y_{n+1} \geq 0. \tag{27.3}$$

Consider now the transformation

$$V_1 = Y_1, V_2 = Y_2, \ldots, V_n = Y_n \text{ and } V_{n+1} = Y_1 + \cdots + Y_{n+1}. \tag{27.4}$$

Then, after noting that the Jacobian of this transformation is 1, we obtain from (27.3) the joint density function of $V_1, V_2, \ldots, V_{n+1}$ as

$$p_{V_1,\ldots,V_{n+1}}(v_1, \ldots, v_{n+1}) = e^{-v_{n+1}}, \quad v_1, \ldots, v_n \geq 0, \quad \sum_{i=1}^{n} v_i \leq v_{n+1}. \tag{27.5}$$

Now, consider the transformation

$$X_1 = \frac{V_1}{V_{n+1}}, \ \ldots, X_n = \frac{V_n}{V_{n+1}} \quad \text{and} \quad X_{n+1} = V_{n+1} \qquad (27.6)$$

or equivalently,

$$V_1 = X_1 X_{n+1}, \ \ldots, V_n = X_n X_{n+1}, \quad \text{and} \quad V_{n+1} = X_{n+1}. \qquad (27.7)$$

From (27.7), it is clear that the Jacobian of this transformation is X_{n+1}^n. Then, we obtain from (27.5) the joint density function of $X_1, X_2, \ldots, X_{n+1}$ as

$$p_{X_1,\ldots,X_{n+1}}(x_1,\ldots,x_{n+1}) = e^{-x_{n+1}} x_{n+1}^n,$$

$$0 \le x_1,\ldots,x_n \le 1, \ 0 \le \sum_{i=1}^{n} x_i \le 1, \ x_{n+1} \ge 0. \qquad (27.8)$$

Integrating out the variable x_{n+1} from (27.8), we then obtain the joint density function of $X_1, \ldots, X_n$ as

$$p_{X_1,\ldots,X_n}(x_1,\ldots,x_n) = n!, \ 0 \le x_1,\ldots,x_n \le 1, \ 0 \le \sum_{i=1}^{n} x_i \le 1. \qquad (27.9)$$

In addition, from the density functions in (27.8) and (27.9), we readily observe that the random vector $(X_1, \ldots, X_n)$ and the random variable X_{n+1} are statistically independent.

We thus have the joint density function of the uniform spacings $T_{1,n}, \ldots, T_{n,n}$ to be given by (27.9). If A_n denotes the region

$$A_n = \left\{ (x_1,\ldots,x_n): \ 0 \le x_1,\ldots,x_n \le 1, \ 0 \le \sum_{i=1}^{n} x_i \le 1 \right\}, \qquad (27.10)$$

then we immediately have from the joint density function in (27.9) that

$$\int \cdots \int_{A_n} dx_1 \cdots dx_n = \frac{1}{n!}. \qquad (27.11)$$

It turns out that (27.11) is the simplest case of the well-known *Dirichlet integral formula*, which, in its general form, is [Dirichlet (1839)]

$$\int \cdots \int_{A_n} x_1^{a_1-1} \cdots x_n^{a_n-1} \left(1 - \sum_{i=1}^{n} x_i \right)^{a_{n+1}-1} dx_1 \cdots dx_n$$

$$= \frac{\Gamma(a_1) \cdots \Gamma(a_{n+1})}{\Gamma(a_1 + \cdots + a_{n+1})} \qquad (27.12)$$

for a_k's positive, where $\Gamma(\cdot)$ denotes the complete gamma function. For details on the history of the Dirichlet integral formula above and its role in probability and statistics, one may refer to Gupta and Richards (2001).

27.2 Derivation of Dirichlet Formula

In deriving the simpler formula in (27.11), we started with the random variables $Y_1, \ldots, Y_{n+1}$ as independent standard exponential random variables. As a natural generalization, let us now assume that $Y_1, \ldots, Y_{n+1}$ are independent gamma random variables with $Y_k \sim \Gamma(a_k, 0, 1)$, where $a_k > 0$. Then, the joint density function of $Y_1, \ldots, Y_{n+1}$ is

$$p_{Y_1,\ldots,Y_{n+1}}(y_1, \ldots, y_{n+1})$$
$$= \frac{1}{\Gamma(a_1) \cdots \Gamma(a_{n+1})} e^{-(y_1+\cdots+y_{n+1})} y_1^{a_1-1} \cdots y_n^{a_n-1},$$
$$y_1, \ldots, y_{n+1} \geq 0. \tag{27.13}$$

With the transformation in (27.4), we obtain the joint density function of $V_1, \ldots, V_{n+1}$ as

$$p_{V_1,\ldots,V_{n+1}}(v_1, \ldots, v_{n+1})$$
$$= \frac{1}{\Gamma(a_1) \cdots \Gamma(a_{n+1})} e^{-v_{n+1}} v_1^{a_1-1} \cdots v_n^{a_n-1} \left(v_{n+1} - \sum_{i=1}^{n} v_i\right)^{a_{n+1}-1},$$
$$v_1, \ldots, v_n \geq 0, \ \sum_{i=1}^{n} v_i \leq v_{n+1}. \tag{27.14}$$

Now making the transformation in (27.7) (with the Jacobian as X_{n+1}^n), we obtain from (27.14) the joint density function of $X_1, \ldots, X_{n+1}$ as

$$p_{X_1,\ldots,X_{n+1}}(x_1, \ldots, x_{n+1})$$
$$= \frac{1}{\Gamma(a_1) \cdots \Gamma(a_{n+1})} e^{-x_{n+1}} x_1^{a_1-1} \cdots x_n^{a_n-1} x_{n+1}^{a_1+\cdots+a_{n+1}-1}$$
$$\times \left(1 - \sum_{i=1}^{n} x_i\right)^{a_{n+1}-1},$$
$$0 \leq x_1, \ldots, x_n \leq 1, \ 0 \leq \sum_{i=1}^{n} x_i \leq 1, \ x_{n+1} \geq 0. \tag{27.15}$$

Upon integrating out the variable x_{n+1} in (27.15), we then obtain the joint density function of $X_1, \ldots, X_n$ as

$$p_{X_1,\ldots,X_n}(x_1, \ldots, x_n)$$
$$= \frac{\Gamma(a_1 + \cdots + a_{n+1})}{\Gamma(a_1) \cdots \Gamma(a_{n+1})} x_1^{a_1-1} \cdots x_n^{a_n-1} \left(1 - \sum_{i=1}^{n} x_i\right)^{a_{n+1}-1},$$
$$0 \leq x_1, \ldots, x_n \leq 1, \ 0 \leq \sum_{i=1}^{n} x_i \leq 1. \tag{27.16}$$

Indeed, from the fact that (27.16) represents a density function, we readily obtain

$$
\int \cdots \int_{A_n} x_1^{a_1-1} \cdots x_n^{a_n-1} \left(1 - \sum_{i=1}^{n} x_i\right)^{a_{n+1}-1} dx_1 \cdots dx_n
$$

$$
= \frac{\Gamma(a_1) \cdots \Gamma(a_{n+1})}{\Gamma(a_1 + \cdots + a_{n+1})},
$$

which is exactly the Dirichlet integral formula presented in (27.12). We thus have a multivariate density function in (27.16) which is very closely related to the multidimensional integral in (27.12) evaluated by Dirichlet (1839).

27.3   Notations

A random vector $\mathbf{X} = (X_1, \ldots, X_n)$ is said to have an n-dimensional *standard Dirichlet distribution* with positive parameters $a_1, \ldots, a_{n+1}$ if its density function is given by (27.16) and is denoted by $\mathbf{X} \sim \mathcal{D}_n(a_1, \ldots, a_{n+1})$. Note that when $n = 1$, (27.16) reduces to

$$
p_X(x) = \frac{\Gamma(a_1 + a_2)}{\Gamma(a_1)\Gamma(a_2)} \, x^{a_1-1}(1 - x)^{a_2-1}, \qquad 0 \le x \le 1, \; a_1, a_2 > 0,
$$

which is nothing but the standard beta distribution discussed in Chapter 16. Indeed, the linear transformation of the random variables $X_1, \ldots, X_n$ will yield the random vector $(b_1 + c_1 X_1, \ldots, b_n + c_n X_n)$ having an n-dimensional general Dirichlet distribution with shape parameter $(a_1, \ldots, a_{n+1})$, location parameters $(b_1, \ldots, b_n)$, and scale parameters $(c_1, \ldots, c_n)$. However, for the rest of this chapter we consider only the standard Dirichlet distribution in (27.16), due to its simplicity.

27.4   Marginal Distributions

Let $\mathbf{X} \sim \mathcal{D}_n(a_1, \ldots, a_{n+1})$. Then, as seen in Section 27.2, the components $X_1, \ldots, X_n$ admit the representations

$$
(X_1, \ldots, X_n) \overset{d}{=} \left(\frac{Y_1}{S_{n+1}}, \ldots, \frac{Y_n}{S_{n+1}}\right) \tag{27.17}
$$

and

$$
X_k \overset{d}{=} \frac{Y_k}{S_{n+1}}, \qquad k = 1, \ldots, n, \tag{27.18}
$$

where $Y_1, \ldots, Y_{n+1}$ are independent standard gamma random variables with $Y_k \sim \Gamma(a_k, 0, 1)$ and $S_{n+1} = Y_1 + \cdots + Y_{n+1}$. From the properties of gamma distributions (see Section 20.7), it is then known that

$$
S_{n+1} \sim \Gamma(a, 0, 1) \quad \text{with} \quad a = a_1 + \cdots + a_{n+1},
$$

$$S_{n+1} - Y_k \sim \Gamma(a - a_k, 0, 1) \text{ (independent of } Y_k)$$

and

$$X_k \overset{d}{=} \frac{Y_k}{S_{n+1}} \overset{d}{=} \frac{Y_k}{Y_k + (S_{n+1} - Y_k)} \sim Be(a_k, a - a_k); \qquad (27.19)$$

that is, the marginal distribution of X_k is beta $Be(a_k, a - a_k), k = 1, \ldots, n$ (note that this is just a Dirichlet distribution with $n = 1$). Thus, the Dirichlet distribution forms a natural multivariate generalization of the beta distribution.

Exercise 27.1 From the density function of $\mathbf{X}$ in (27.16), show by means of direct integration that the marginal distribution of X_k is $Be(a_k, a - a_k)$.

For two-dimensional marginal distributions, it follows from (27.17) that for any $1 \le k < \ell \le n$,

$$(X_k, X_\ell) \overset{d}{=} \left(\frac{Y_k}{Y_k + Y_\ell + Z}, \frac{Y_\ell}{Y_k + Y_\ell + Z} \right), \qquad (27.20)$$

where $Y_k \sim \Gamma(a_k, 0, 1), Y_\ell \sim \Gamma(a_\ell, 0, 1), Z = S_{n+1} - Y_k - Y_\ell \sim \Gamma(a - a_k - a_\ell, 0, 1)$, and Y_k, Y_ℓ, and Z are independent. Thus, (27.20) readily implies that

$$(X_k, X_\ell) \overset{d}{=} \mathcal{D}_2(a_k, a_\ell, a - a_k - a_\ell). \qquad (27.21)$$

In a similar manner, we find that

$$\left(X_{k(1)}, \ldots, X_{k(r)} \right) \sim \mathcal{D}_r \left(a_{k(1)}, \ldots, a_{k(r)}, a - \sum_{i=1}^{r} a_{k(i)} \right) \qquad (27.22)$$

for any $r = 1, \ldots, n$ and $1 \le k(1) < k(2) < \cdots < k(r) \le n$.

Exercise 27.2 If $(X_1, \ldots, X_6) \sim \mathcal{D}_6(a_1, \ldots, a_7)$, show that

$$(X_1, X_2 + X_3, X_4 + X_5 + X_6) \sim \mathcal{D}_3(a_1, a_2 + a_3, a_4 + a_5 + a_6, a_7).$$

Exercise 27.3 Let $(X_1, \ldots, X_n) \sim \mathcal{D}_n(a_1, \ldots, a_{n+1})$. Find the distribution of $W_k = X_1 + \cdots + X_k$.

Exercise 27.4 Let $(X_1, X_2) \sim \mathcal{D}_2(a_1, a_2, a_3)$. Obtain the conditional density function of X_1, given $X_2 = x_2$. Do you observe some connection to the beta distribution? Derive an expression for the conditional mean $E(X_1|X_2 = x_2)$ and comment.

27.5   Marginal Moments

Let $\mathbf{X} \sim \mathcal{D}_n(a_1, \ldots, a_{n+1})$. In this case, as shown in Section 27.4, the marginal distribution of X_k is beta $Be(a_k, a - a_k)$, where $a = a_1 + \cdots + a_{n+1}$. Then, from the formulas of moments of beta distribution presented in Section 16.5, we immediately have

$$
EX_k = \frac{a_k}{a},
$$

$$
EX_k^2 = \frac{a_k(a_k + 1)}{a(a + 1)},
$$

$$
\mathrm{Var}\, X_k = \frac{a_k(a - a_k)}{a^2(a + 1)}.
$$

27.6   Product Moments

Let $\mathbf{X} \sim \mathcal{D}_n(a_1, \ldots, a_{n+1})$. Then, from the density function in (27.16), we have the product moment of order $(\alpha_1, \ldots, \alpha_n)$ as

$$
\begin{aligned}
&E\left(X_1^{\alpha_1} \cdots X_n^{\alpha_n}\right) \\
&= \int \cdots \int_{A_n} x_1^{\alpha_1} \cdots x_n^{\alpha_n}\, p_{X_1, \ldots, X_n}(x_1, \ldots, x_n)\, dx_1 \cdots dx_n \\
&= \frac{\Gamma(a_1 + \cdots + a_{n+1})}{\Gamma(a_1) \cdots \Gamma(a_{n+1})} \int \cdots \int_{A_n} x_1^{a_1 + \alpha_1 - 1} \cdots x_n^{a_n + \alpha_n - 1} \\
&\qquad \times \left(1 - \sum_{i=1}^{n} x_i\right)^{a_{n+1} - 1} dx_1 \cdots dx_n \\
&= \frac{\Gamma(a_1 + \cdots + a_{n+1})}{\Gamma(a_1 + \cdots + a_{n+1} + \alpha_1 + \cdots + \alpha_n)} \prod_{i=1}^{n} \frac{\Gamma(a_i + \alpha_i)}{\Gamma(a_i)}, \quad (27.23)
\end{aligned}
$$

where the last equality follows by an application of the Dirichlet integral formula in (27.12).

In particular, we obtain from (27.23) that

$$
E(X_1 \cdots X_n) = \frac{\Gamma(a_1 + \cdots + a_{n+1})}{\Gamma(a_1 + \cdots + a_{n+1} + n)}\, a_1 a_2 \cdots a_n = \prod_{i=1}^{n} \frac{a_i}{a + i - 1}
$$

and

$$
E(X_k X_\ell) = \frac{\Gamma(a_1 + \cdots + a_{n+1})}{\Gamma(a_1 + \cdots + a_{n+1} + 2)}\, a_k a_\ell = \frac{a_k a_\ell}{a(a + 1)} \quad \text{for } k \neq \ell.
$$

Exercise 27.5 Let $\mathbf{X} \sim \mathcal{D}_n(a_1, \ldots, a_{n+1})$. Derive the covariance and correlation coefficient between X_k and X_ℓ.

27.7   Dirichlet Distribution of Second Kind

In Chapter 16, when dealing with a beta random variable $X \sim Be(a,b)$ with probability density function

$$p_X(x) = \frac{\Gamma(a+b)}{\Gamma(a)\Gamma(b)}\, x^{a-1}(1-x)^{b-1},\ 0 < x < 1,\ a,b > 0,$$

by considering the transformation $Y = X/(1-X)$ or equivalently $X = Y/(1+Y)$, we introduced the beta distribution of the second kind with probability density function

$$p_Y(y) = \frac{\Gamma(a+b)}{\Gamma(a)\Gamma(b)} \cdot \frac{y^{a-1}}{(1+y)^{a+b}},\qquad y > 0,\ a,b > 0.$$

In a similar manner, we shall now introduce the *Dirichlet distribution of the second kind*. Specifically, let $\mathbf{X} \sim \mathcal{D}_n(a_1,\ldots,a_{n+1})$, where $a_k > 0$ ($k = 1,\ldots,n+1$) are positive parameters. Now, consider the transformation

$$Y_1 = \frac{X_1}{1 - X_1 - \cdots - X_n},\quad \cdots\ , Y_n = \frac{X_n}{1 - X_1 - \cdots - X_n} \qquad (27.24)$$

or equivalently,

$$X_1 = \frac{Y_1}{1 + Y_1 + \cdots + Y_n},\quad \cdots\ , X_n = \frac{Y_n}{1 + Y_1 + \cdots + Y_n}. \qquad (27.25)$$

Then, it can be shown that the Jacobian of this transformation is $(1 + Y_1 + \cdots + Y_n)^{-(n+1)}$. Then, we readily obtain from (27.16) the density function of $\mathbf{Y} = (Y_1,\ldots,Y_n)$ as

$$p_{Y_1,\ldots,Y_n}(y_1,\ldots,y_n)$$
$$= \frac{\Gamma(a_1 + \cdots + a_{n+1})}{\Gamma(a_1)\cdots\Gamma(a_{n+1})} \cdot \frac{y_1^{a_1-1}\cdots y_n^{a_n-1}}{(1 + y_1 + \cdots + y_n)^{a_1+\cdots+a_{n+1}}},$$
$$y_1,\ldots,y_n > 0,\ a_1,\ldots,a_{n+1} > 0. \qquad (27.26)$$

The density function (27.26) is the Dirichlet density of the second kind.

Exercise 27.6 Show that the Jacobian of the transformation in (27.25) is $(1 + Y_1 + \cdots + Y_n)^{-(n+1)}$ (use elementary row and column operations).

Exercise 27.7 Suppose that $\mathbf{Y}$ has a Dirichlet distribution of the second kind in (27.26). Derive explicit expressions for EY_k, Var Y_k, Cov(Y_k,Y_ℓ), and correlation $\rho(Y_k,Y_\ell)$.

27.8   Liouville Distribution

Liouville (1839) generalized the Dirichlet integral formula in (27.12) by establishing that

$$
\int \cdots \int_{x_1+\cdots+x_n<h} f(x_1 + \cdots + x_n) \, x_1^{a_1-1} \cdots x_n^{a_n-1} \, dx_1 \cdots dx_n
$$

$$
= \frac{\Gamma(a_1)\cdots\Gamma(a_n)}{\Gamma(a_1+\cdots+a_n)} \int_0^h f(t) t^{a_1+\cdots+a_n-1} \, dt, \qquad (27.27)
$$

where $a_1,\ldots,a_n$ are positive parameters, $x_1,\ldots,x_n$ are positive, and $f(\cdot)$ is a suitably chosen function. It is clear that if we set $h=1$ and choose $f(t)=(1-t)^{a_{n+1}-1}$, (27.27) readily reduces to the Dirichlet integral formula in (27.12). Also, by letting $h\to\infty$ in (27.27), we obtain the *Liouville integral formula*

$$
\int_0^\infty \cdots \int_0^\infty f(x_1 + \cdots + x_n) \, x_1^{a_1-1} \cdots x_n^{a_n-1} \, dx_1 \cdots dx_n
$$

$$
= \frac{\Gamma(a_1)\cdots\Gamma(a_n)}{\Gamma(a_1+\cdots+a_n)} \int_0^\infty f(t) t^{a_1+\cdots+a_n-1} \, dt, \qquad (27.28)
$$

where $a_1,\ldots,a_n>0$ and $t^{a_1+\cdots+a_n-1}f(t)$ is integrable on $(0,\infty)$.

The Liouville integral formula in (27.28) readily yields *Liouville distribution* with probability density function

$$
p_{X_1,\ldots,X_n}(x_1,\ldots,x_n) = C \, f(x_1 + \cdots + x_n) \, x_1^{a_1-1} \cdots x_n^{a_n-1},
$$

$$
x_1,\ldots,x_n>0, \ a_1,\ldots,a_n>0, \qquad (27.29)
$$

where C is a normalizing constant and $f(\cdot)$ is a nonnegative function such that $f(t)t^{a_1+\cdots+a_n-1}$ is integrable on $(0,\infty)$. For a historical view and details on the Liouville distribution, one may refer to Gupta and Richards (2001).

Exercise 27.8 Show that the Dirichlet distribution of the second kind in (27.26) is a Liouville distribution by choosing the function $f(t)$ appropriately and then determining the constant C from the Liouville integral formula in (27.28).

APPENDIX

PIONEERS IN DISTRIBUTION THEORY

As is evident from the preceding chapters, several prominent mathematicians and statisticians have made pioneering contributions to the area of *statistical distribution theory*. To give students a historical sense of developments in this important and fundamental area of statistics, we present here a brief biographical sketch of these major contributors.

Bernoulli, Jakob

Born – January 6,1655, in Basel, Switzerland
Died – August 16, 1705, in Basel, Switzerland

Jakob Bernoulli was the first of the Bernoulli family of Swiss mathematicians. His work *Ars Conjectandi (The Art of Conjecturing)*, published posthumously in 1713 by his nephew N. Bernoulli, contained the Bernoulli law of large numbers for Bernoulli sequences of independent trials. Usually, a random variable, taking values 1 and 0 with probabilities p and $1 - p$, $0 \leq p \leq 1$, is said to have the *Bernoulli distribution*. Sometimes, the binomial distributions, which are convolutions of Bernoulli distributions, are also called the Bernoulli distributions.

Burr, Irving W.

Born – April 9, 1908, in Fallon, Nevada, United States
Died – March 13, 1989, in Sequim, Washington, United States

Burr, in a famous paper in 1942, proposed a number of forms of explicit cumulative distribution functions that might be useful for purposes of graduation. There were 12 different forms presented in the paper, which have since come to be known as the *Burr system of distributions*, have been studied quite extensively in the literature. A number of well-known distributions such as the uniform, Rayleigh, logistic, and log-logistic are present in Burr's system as special cases. In the years following, Burr worked on inferential problems and fitting methods for some of these forms of distributions. In one of his last papers (co-authored with E. S. Pearson and N. L. Johnson), he also made an extensive comparison of different systems of frequency curves.

Cauchy, Augustin-Louis

Born – August 21, 1789, in Paris, France
Died – May 23, 1857, in Sceaux (near Paris), France

Augustin-Louis Cauchy was a renowned French mathematician. He investigated the so-called Cauchy functions $p(x, h, \alpha)$, which had Fourier transformations of the form

$$f(t) = \exp\left(-h|t|^{\alpha}\right), \qquad h > 0, \ \alpha > 0.$$

It was proved later that $p(x, h, \alpha)$, $0 < \alpha \leq 2$, are indeed probability density functions. The special case of the distribution with pdf

$$p(x, 1, 1) = \frac{1}{\pi(1 + x^2)}$$

and with characteristic function

$$f(t) = \exp(-|t|)$$

is called the standard *Cauchy distribution*. The general Cauchy distribution, of course, has the pdf $p((x - a), h, 1)$.

Dirichlet, Johann Peter Gustav Lejeune

Born – February 13, 1805, in Düren, French Empire (now Germany)
Died – May 5, 1859, in Göttingen, Hanover, Germany

Lejeune Dirichlet's family came from the neighborhood of Liège in Belgium and not, as many had claimed, from France. Dirichlet had some of the renowned mathematicians as teachers and profited greatly from his contacts with Biot, Fourier, Hachette, Laplace, Legendre, and Poisson. Dirichlet made pioneering contributions to different areas of mathematics starting with his famous first paper on *Fermat's last theorem*. In 1839, Dirichlet established the general integral formula

$$\int \cdots \int_{A_n} \left(1 - \sum_{i=1}^{n} u_i\right)^{a_{n+1}-1} \prod_{i=1}^{n} u_i^{a_i-1} \, du_i = \frac{\prod_{i=1}^{n+1} \Gamma(a_i)}{\Gamma\left(\sum_{i=1}^{n+1} a_i\right)},$$

where

$$A_n = \left\{(u_1, \ldots, u_n) : \ 0 \leq u_1, \ldots, u_n \leq 1, \ \sum_{i=1}^{n} u_i \leq 1\right\}.$$

The above n-dimensional integral is now known as the *Dirichlet integral* and the probability distribution arising from the integral formula as the *Dirichlet distribution*.

Fisher, Ronald Aylmer

Born – February 17, 1890, in London, England
Died – July 29, 1962, in Adelaide, Australia

Fisher was a renowned British statistician and geneticist and is generally regarded as the founder of the field of statistics. The *Fisher's F-distribution*, having the pdf

$$p_{\alpha,\beta}(x) = \frac{1}{B(\alpha/2, \beta/2)} \left(\frac{\alpha}{\beta}\right)^{\alpha/2} x^{(\beta/2)-1} \left(1 + \frac{\alpha}{\beta} x\right)^{-(\alpha+\beta)/2}, \quad x, \ \alpha, \ \beta > 0,$$

plays a very important role in many statistical inferential problems. One more distribution used in the analysis of variance, called the *Fisher's z-distribution*, has the pdf

$$p_{m,n}(x) = 2m^{m/2}n^{n/2} \frac{\Gamma\left(\frac{m+n}{2}\right) e^{mx}}{\Gamma(m/2)\Gamma(n/2)(me^{2x}+n)^{(m+n)/2}}, \quad -\infty < x < \infty,$$

where $m, n = 1, 2, \ldots$.

Fréchet, René-Maurice

Born – September 2, 1878, in Maligny, France
Died – June 4, 1973, in Paris, France

Fréchet was a renowned French mathematician who successfully combined his work in the areas of topology and the theory of abstract spaces to make his essential contribution to statistics. In 1927, he derived one of the three possible limiting distributions for extremes. Hence, the family of distributions with cdf

$$G(x, \alpha) = \exp(-x^{-\alpha}), \qquad x > 0, \ \alpha > 0,$$

is sometimes referred to as the *Fréchet type of distributions*.

Gauss, Carl Friedrich

Born – April 30, 1777, in Braunschweig, Duchy of Brunswick, Germany
Died – February 23, 1855, in Göttingen, Hanover, Germany

Gauss is undisputably one of the greatest mathematicians of all time. In his theory of errors in 1809, Gauss suggested that the normal distributions for erros with density

$$\frac{h}{\sqrt{\pi}} \exp\{-h^2 x^2\}, \qquad -\infty < x < \infty, \ h > 0.$$

For this reason, the normal distribution is often called the *Gaussian law* or as the *Gaussian distribution*.

Gnedenko, Boris Vladimirovich

Born – January 1, 1912, in Simbirsk (Ulyanovsk), Russia
Died – December 27, 1995, in Moscow, Russia

In the first half of the twentieth century, foundations for the theory of extreme values were laid by R.-M. Fréchet, R. von Mises, R. A. Fisher, and L. H. C. Tippett. Consolidating these works with his own, Gnedenko produced his outstanding paper in 1943 that discussed at great length the asymptotic behavior of extremes for independent and identically distributed random variables. For this reason, sometimes the limiting distributions of maximal values are called *three types of Gnedenko's limiting distributions.*

Gosset, William Sealey

Born – June 13, 1876, in Canterbury, England
Died – October 16, 1937, in Beaconsfield, England

Being a chemist and later a successful statistical assistant in the Guinness brewery, Gosset did important early work on statistics, which he wrote under the pseudonym *Student.* In 1908, he proposed the use of the t-test for quality control purposes in brewing industry. The corresponding distribution has pdf

$$p(t) = \frac{\Gamma\left(\dfrac{n+1}{2}\right)}{\sqrt{n\pi}\,\Gamma\left(n/2\right)} \left(1 + \frac{t^2}{n}\right)^{-(n+1)/2}, \qquad -\infty < t < \infty,$$

and it is called *Student's t distribution with n degrees of freedom.*

Helmert, Friedrich Robert

Born – July 31, 1843, in Freiberg (Saxony), Germany
Died – June 15, 1917, in Potsdam (Prussia), Germany

Helmert was a famous German mathematical physicist whose main research was in geodesy, which led him to investigate several statistical problems. In a famous paper in 1876, Helmert first proved, for a random sample $X_1, \ldots, X_n$ from a normal $N(a, \sigma^2)$ distribution, the independence of $\bar{X}$ and any function of $X_1 - \bar{X}, \ldots, X_n - \bar{X}$, including the variable $S^2 = \sum_{j=1}^n (X_j - \bar{X})^2 / \sigma^2$. Then, using a very interesting transformation of variables, which is now referred to in the literature as *Helmert's transformation*, he proved that S^2 is distributed as chi-square with $n - 1$ degrees of freedom.

Johnson, Norman Lloyd

Born – January 9, 1917, in Ilford, Essex, England

Johnson, having started his statistical career in London in 1940s, came into close contact and collaboration with such eminent statisticians as B. L.

Welch, Egon Pearson, and F. N. David. Motivated by the Pearson family of distributions and the idea that it would be very convenient to have a family of distributions, produced by a simple transformation of normal variables, such that for any pair of values $(\sqrt{\gamma_1}, \gamma_2)$ there is just one member of this family of distributions, Johnson in 1949 proposed a collection of three transformations. These resulted in *lognormal*, S_B, and S_U distributions, which are now referred to as *Johnson's system of distributions*. In addition, his book *System of Frequency Curves* (co-authored with W. P. Elderton), published by Cambridge University Press, and the series of four volumes on *Distributions in Statistics* (co-authored with S. Kotz), published by John Wiley & Sons and revised, have become classic references to anyone working on statistical distributions and their applications.

Kolmogorov, Andrey Nikolayevich

Born – April 25, 1903, in Tambov, Russia
Died – October 20, 1987, in Moscow, Russia

Kolmogorov, one of the most outstanding mathematicians of the twentieth century, produced many remarkable results in different fields of mathematics. His deep work initiated developments on many new directions in modern probability theory and its applications. In 1933, he proposed one of the most popular goodness-of-fit tests called the *Kolmogorov–Smirnov test*. From this test procedure, a new distribution with cdf

$$F(x) = \sum_{n=-\infty}^{\infty} (-1)^n \exp(-2n^2 x^2), \qquad x > 0.$$

originated, which is called the *Kolmogorov distribution* in the literature.

Kotz, Samuel

Born – August 28, 1930, in Harbin, China

Kotz has made significant contributions to many areas of statistics, most notably in the area of statistical distribution theory. His four volumes on *Distributions in Statistics* (co-authored with N. L. Johnson), published by John Wiley & Sons in 1970 and revised, have become classic references to anyone working on statistical distributions and their applications. His 1978 monograph *Characterization of Probability Distributions* (co-authored with J. Galambos), published by Springer-Verlag, and the 1989 book *Multivariate Symmetric Distributions* (co-authored with K. T. Fang and K. W. Ng), published by Chapman & Hall, have become important sources for researchers working on characterization problems and multivariate distribution theory. A family of elliptically symmetric distributions, which includes the multivariate normal distribution as a special case, are known by the name *Kotz–type elliptical distributions* in the literature.

Laplace, Pierre-Simon

Born – March 23, 1749, in Beaumount-en-Auge, France
Died – March 5, 1827, in Paris, France

Laplace was a renowned French astronomer, mathematician, and physicist. In 1812, he published his famous work *Théorie analytique des probabilités* (*Analytic Theory of Probability*), wherein he rationalized the necessity to consider and investigate two statistical distributions, both of which carry his name. The first one is the distribution with pdf

$$p(x) = \frac{1}{2\lambda} \exp\left\{-\frac{|x-a|}{\lambda}\right\}, \qquad -\infty < x < \infty,$$

where $-\infty < a < \infty$ and $\lambda > 0$, which is called the *Laplace distribution* and sometimes the *first law of Laplace*. The second one is the normal, which is sometimes called the *second law of Laplace* or *Gauss–Laplace distribution*.

Linnik, Yuri Vladimirovich

Born – January 21, 1915, in Belaya Tserkov, Russia (Ukraine)
Died – June 30, 1972 , in Leningrad (St. Petersburg), Russia

Linnik was the founder of the modern St. Petersburg (Leningrad) school of probability and mathematical statistics. He was the first to prove that

$$f_\alpha(t) = (1 + |t|^\alpha)^{-1}$$

(for any $0 < \alpha \leq 2$) is indeed a characteristic function of some random variable. For this reason, distributions with characteristic functions $f_\alpha(t)$ are known in the literature as the *Linnik distributions*. His famous book *Characterization Problems of Mathematical Statistics* (co-authored with A. Kagan and C. R. Rao), published in 1973 by John Wiley & Sons, also became a basic source of reference, inspiration, and ideas for many involved with research on characterizations of probability distributions.

Maxwell, James Clerk

Born – June 13, 1831, in Edinburgh, Scotland
Died – November 5, 1879, in Cambridge, England

James Maxwell was a famous Scottish physicist who is regarded as the founder of classical thermodynamics. In 1859, Maxwell was first to suggest that the velocities of molecules in a gas, previously assumed to be equal, must follow the chi distribution with three degrees of freedom having the density

$$p(x) = \sqrt{2/\pi}\, x^2\, \exp\left\{-x^2/2\right\}, \qquad x > 0.$$

This distribution, therefore, is aptly called the *Maxwell distribution*.

Pareto, Vilfredo

Born – July 15, 1848, in Paris, France
Died – August 20, 1923, in Geneva, Switzerland

Pareto was a renowned Italian economist and sociologist. He was one of the first who tried to explain and solve economic problems with the help of statistics. In 1897, he formulated his law of income distributions, where cdf's of the form

$$F(x) = 1 - \left(\frac{h}{x}\right)^{\alpha}, \qquad x > h, \ \alpha > 0$$

played a very important role. For this reason, these distributions are referred to in the literature as the *Pareto distributions*.

Pascal, Blaise

Born – June 19, 1623, in Clermont-Ferrand, France
Died – August 19, 1662, in Paris, France

Pascal was a famous French mathematician, physicist, religious philosopher, and one of the founders of probability theory. A discrete random variable taking on values $0, 1, \ldots$ with probabilities

$$p_n = \binom{r+n-1}{r-1} p^r (1-p)^n, \qquad n = 0, 1, 2, \ldots,$$

where $0 < p < 1$ and r is a positive integer, is said to have the *Pascal distribution* with parameters p and r. This distribution is, of course, a special case of the negative binomial distributions.

Pearson, Egon Sharpe

Born – August 11, 1895, in Hampstead, London, England
Died – June 12, 1980, in Midhurst, Sussex, England

Egon Pearson, the only son of the eminent British statistician Karl Pearson, was influenced by his father in his early academic career and later by his correspondence and association with "Student" (W. S. Gosset) and Jerzy Neyman. His collaboration with Neyman resulted in the now-famous Neyman–Pearson approach to hypothesis testing. He successfully revised Karl Pearson's *Tables for Statisticians and Biometricians* jointly with L. J. Comrie, and later with H. O. Hartley producing *Biometrika Tables for Statisticians*. Egon Pearson constructed many important statistical tables, including those of percentage points of Pearson curves and distribution of skewness and kurtosis coefficients. Right up to his death, he continued his work on statistical distributions, and, in fact, his last paper (co-authored with N. L. Johnson and I. W. Burr) dealt with a comparison of different systems of frequency curves.

Pearson, Karl

Born – March 27, 1857, in London, England
Died – April 27, 1936, in Coldharbour, Surrey, England

Karl Pearson, an eminent British statistician, is regarded as the founding father of statistical distribution theory. Inspired by the famous book *Natural Inheritance* of Francis Galton published in 1889, he started his work on evolution by examining large data sets (collected by F. Galton) when he noted systematic departures from normality in most cases. This led him to the development of the *Pearson system of frequency curves* in 1895. In addition, he also prepared *Tables for Statisticians and Biometricians* in order to facilitate the use of statistical methods by practitioners. With the support of F. Galton and W. F. R. Weldon, he founded the now-prestigious journal *Biometrika* in 1901, which he also edited from its inception until his death in 1936.

Poisson, Siméon-Denis

Born – June 21, 1781, in Pithiviers, France
Died – April 25, 1840, in Sceaux (near Paris), Paris

Poisson was a renowned French mathematician who made fundamental work in applications of mathematics to problems in electricity, magnetism, and mechanics. In 1837, in his work *Recherches sur la probabilité des jugements en matière criminelle et en matière civile, précédées des règles générales du calcul des probabilités*, the Poisson distribution first appeared as an approximation to the binomial distribution. A random variable has the *Poisson distribution* if it takes on values $0, 1, \ldots$ with probabilities

$$p_n = \frac{e^{-\lambda} \lambda^n}{n!}, \qquad n = 0, 1, \ldots, \ \lambda > 0.$$

Pólya, George

Born – December 13, 1887, in Budapest, Hungary
Died – September 7, 1985, in Palo Alto, California, United States

Pólya was a famous Hungarian-born U.S. mathematician who is known for his significant contributions to combinatorics, number theory, and probability theory. He was also an author of some popular books on the problem-solving process, such as *How to Solve It* and *Mathematical Discovery*. In 1923, Pólya discussed a special urn model which generated a probability distribution with probabilities

$$p_k = \binom{n}{k} \frac{p(p+\alpha)\cdots\{p+(k-1)\alpha\}q(q+\alpha)\cdots\{q+(n-k-1)\alpha\}}{(1+\alpha)(1+2\alpha)\cdots\{1+(n-1)\alpha\}},$$

corresponding to the values $0, 1, \ldots, n$, where $0 < p < 1$, $q = 1 - p$, $\alpha > 0$, and $n = 1, 2, \ldots$. For this reason, this distribution is called as the *Pólya distribution*.

Rao, Calyampudi Radhakrishna

Born – September 10, 1920, in Huvvinna Hadagalli (Karnataka), India

Rao is considered to be on the most creative thinkers in the field of statistics and one of the few pioneers who brought statistical theory to its maturity in the twentieth century. He has made numerous fundamental contributions to different areas of statistics. The *Rao–Rubin characterization* of the Poisson distribution and the Lau–Rao characterization theorem based on integrated Cauchy functional equation, published in 1964 and 1982, generated great interest in characterizations of statistical distributions. Another significant contribution he made in the area of statistical distribution theory is on weighted distributions. In addition, his books *Characterization Problems of Mathematical Statistics* (co-authored with A. Kagan and Yu. V. Linnik), published in 1973 by John Wiley & Sons, and *Choquet–Deny Type Functional Equations with Applications to Stochastic Models* (co-authored with D. N. Shanbhag), published in 1994 by John Wiley & Sons, have also become a basic source of reference, inspiration, and ideas for many involved with research on characterizations of probability distributions.

Rayleigh, 3rd Baron (Strutt, John William)

Born – November 12,1842, in Langford Grove (near Maldon), England
Died – June 30, 1919, in Terling Place (near Witham), England

Lord Rayleigh was a famous English physicist who was awarded the Nobel Prize for Physics in 1904 for his discoveries in the fields of acoustics and optics. In 1880, in his work connected with the wave theory of light, Rayleigh considered a distribution with pdf of the form

$$p(x) = \frac{x}{h^2} \, e^{-x^2/(2h^2)}, \qquad x > 0, \ h > 0.$$

This distribution is aptly called the *Rayleigh distribution* in the literature.

Snedecor, George Waddel

Born – October 20, 1881, in Memphis, Tennessee, United States
Died – February 15, 1974, in Amherst, Massachusetts, United States

Snedecor was a famous U.S. statistician and was the first director of the Statistical Laboratory at Iowa State University, the first of its kind in the United States. In 1948, he also served as the president of the American Statistical Association. The Fisher F-distribution (mentioned earlier) having the pdf

$$p_{\alpha,\beta}(x) = \frac{1}{B\left(\alpha/2, \beta/2\right)} \left(\frac{\alpha}{\beta}\right)^{\alpha/2} x^{(\beta/2)-1} \left(1 + \frac{\alpha}{\beta} \, x\right)^{-(\alpha+\beta)/2},$$
$$x > 0, \ \alpha > 0, \ \beta > 0,$$

is sometimes called as the *Snedecor distribution* or the *Fisher–Snedecor distribution*. In 1937, this distribution was tabulated by Snedecor.

Student (*see* Gosset, William Sealey)

Tukey, John Wilder

> Born – July 16, 1915, in New Bedford, Massachusetts, United States
> Died – July 26, 2000, in Princeton, New Jersey, United States

Tukey was a famous American statistician who made pioneering contributions to many different areas of statistics, most notably on robust inference. To facilitate computationally easy Monte Carlo evaluation of the robustness properties of normal-based inferential procedures, Tukey in 1962 proposed the transformation

$$Y = \begin{cases} \dfrac{X^\lambda - (1-X)^\lambda}{\lambda} & \text{if } \lambda \neq 0 \\ \log\left(\dfrac{X}{1-X}\right) & \text{if } \lambda \to 0, \end{cases}$$

where the random variable X has a standard uniform $U(0,1)$ distribution. For different choices of the shape parameter λ, the transformation above produces a light-tailed or heavy-tailed distributions for the random variable X in addition to providing very good approximations for normal and t distributions. The distributions of X have come to be known as *Tukey's (symmetric) lambda distributions*.

Weibull, Waloddi

> Born – June 18, 1887, in Schleswig-Holstein, Sweden
> Died – October 12, 1979, in Annecy, France

Waloddi Weibull was a famous Swedish engineer and physicist who, in 1939, used distributions of the form

$$G_\alpha(x) = \begin{cases} 0, & x < 0, \\ 1 - \exp\left\{-\left(\dfrac{x-a}{\lambda}\right)^\alpha\right\}, & x > 0, \ \alpha > 0, \\ & -\infty < a < \infty, \ \lambda > 0, \end{cases}$$

to represent the distribution of the breaking strength of materials. In fact, these are limiting distributions of minimal values. In 1951, Weibull also demonstrated a close agreement between many different sets of data and those predicted with the fitted Weibull model, with the data sets used in this study relating to as diverse characteristics as the strength of Bofors' steel, fiber strength of Indian cotton, length of syrtoideas, fatigue life of an ST-37 steel, statures of adult males born in the British Isles, and breadth of the beans *Phaseolus vulgaris*. For this reason, the distributions presented

above are called *Weibull distributions*, and they have become the most popular and commonly used statistical models for lifetime data. Sometimes, the
distributions of maximal values with cdf's

$$H_\alpha(x) = 1 - G_\alpha(-x) = \begin{cases} \exp\left\{-\left(-\dfrac{x-a}{\lambda}\right)^\alpha\right\}, & x < 0, \ \alpha > 0, \\[2mm] & \qquad\qquad -\infty < a < \infty, \ \lambda > 0, \\[1mm] 1, & x > 0, \end{cases}$$

are also called *Weibull-type distributions*.

BIBLIOGRAPHY

Aitchison, J. and Brown, J. A. C. (1957). *The Lognormal Distribution*, Cambridge University Press, Cambridge, England.

Arnold, B. C. (1983). *Pareto Distributions*, International Co-operative Publishing House, Fairland, MD.

Arnold, B. C., Castillo, E. and Sarabia, J.-M. (1999). *Conditionally Specified Distributions*, Springer-Verlag, New York.

Balakrishnan, N. (Ed.) (1992). *Handbook of the Logistic Distribution*, Marcel Dekker, New York.

Balakrishnan, N. and Basu, A. P. (Eds.) (1995). *The Exponential Distribution: Theory, Methods and Applications*, Gordon & Breach Science Publishers, Newark, NJ.

Balakrishnan, N. and Koutras, M. V. (2002). *Runs and Scans with Applications*, John Wiley & Sons, New York.

Bansal, N., Hamedani, G. G., Key, E. S., Volkmer, H., Zhang, H. and Behboodian, J. (1999). Some characterizations of the normal distribution, *Statistics & Probability Letters*, **42**, 393–400.

Bernstein, S. N. (1941). Sur une propriété charactéristique de la loi de Gauss, *Transactions of the Leningrad Polytechnical Institute*, **3**, 21–22.

Box, G. E. P. and Muller, M. E. (1958). A note on the generation of random normal deviates, *Annals of Mathematical Statistics*, **29**, 610–611.

Burr, I. W. (1942). Cumulative frequency functions, *Annals of Mathematical Statistics*, **13**, 215–232.

Cacoullos, T. (1965). A relation between t and F distributions, *Journal of the American Statistical Association*, **60**, 528–531. Correction, **60**, 1249.

Chhikara, R. S. and Folks, J. L. (1989). *The Inverse Gaussian Distribution: Theory, Methodology and Applications*, Marcel Dekker, New York.

Consul, P. C. (1989). *Generalized Poisson Distributions: Properties and Applications*, Marcel Dekker, New York.

Cramér, H. (1936). Über eine Eigenschaft der normalen Verteilungsfunktion, *Mathematische Zeitschrift*, **41**, 405–414.

Crow, E. L. and Shimizu, K. (Eds.) (1988). *Lognormal Distributions: Theory and Applications*, Marcel Dekker, New York.

Darmois, D. (1951). Sur diverses propriétés charactéristique de la loi de probabilité de Laplace-Gauss, *Bulletin of the International Statistical Institute*, **23**(II), 79–82.

Devroye, L. (1990). A note on Linnik's distribution, *Statistics & Probability Letters*, **9**, 305–306.

Dirichlet, J. P. G. L. (1839). Sur une nouvelle méthode pour la détermination des intégrales multiples, *Liouville, Journal de Mathématiques, Series I*, **4**, 164–168. Reprinted in *Dirichlet's Werke* (Eds., L. Kronecker and L. Fuchs), 1969, Vol. 1, pp. 377–380, Chelsea, New York.

Douglas, J. B. (1980). *Analysis with Standard Contagious Distributions*, International Co-operative Publishing House, Burtonsville, MD.

Eggenberger, F. and Pólya, G. (1923). Über die Statistik verketetter Vorgänge, *Zeitschrift für Angewandte Mathematik und Mechanik*, **1**, 279–289.

Evans, M., Peacock, B. and Hastings, N. (2000). *Statistical Distributions*, 3rd ed., John Wiley & Sons, New York.

Feller, W. (1943). On a general class of "contagious" distributions, *Annals of Mathematical Statistics*, **14**, 389–400.

Feller, W. (1968). *An Introduction to Probability Theory and Its Applications*, Vol. 1, 3rd ed., John Wiley & Sons, New York.

Gilchrist, W. G. (2000). *Statistical Modelling with Quantile Functions*, Chapman & Hall/CRC, London, England.

Godambe, A. V. and Patil, G. P. (1975). Some characterizations involving additivity and infinite divisibility and their applications to Poisson mixtures and Poisson sums, In *Statistical Distributions in Scientific Work – 3: Characterizations and Applications* (Eds., G. P. Patil, S. Kotz and J. K. Ord), pp. 339–351, Reidel, Dordrecht.

Govindarajulu, Z. (1963). Relationships among moments of order statistics in samples from two related populations, *Technometrics*, **5**, 514–518.

Gradshteyn, I. S. and Ryzhik, I. M. (Eds.) (1994). *Table of Integrals, Series, and Products*, 5th ed., Academic Press, San Diego, CA.

Gupta, R. D. and Richards, D. St. P. (2001). The history of the Dirichlet and Liouville distributions, *International Statistical Review*, **69**, 433–446.

Haight, F. A. (1967). *Handbook of the Poisson Distribution*, John Wiley & Sons, New York.

Helmert, F. R. (1876). Die Genauigkeit der Formel von Peters zue Berechnung des wahrscheinlichen Beobachtungsfehlers directer Beobachtungen gleicher Genauigkeit, *Astronomische Nachrichten*, **88**, columns 113–120.

Johnson, N. L. (1949). Systems of frequency curves generated by methods of translation, *Biometrika*, **36**, 149–176.

Johnson, N. L. and Kotz, S. (1990). Use of moments in deriving distributions and some characterizations, *The Mathematical Scientist*, **15**, 42–52.

Johnson, N. L., Kotz, S. and Balakrishnan, N. (1994). *Continuous Univariate Distributions*, Vol. 1, 2nd ed., John Wiley & Sons, New York.

Johnson, N. L., Kotz, S. and Balakrishnan, N. (1995). *Continuous Univariate Distributions*, Vol. 2, 2nd ed., John Wiley & Sons, New York.

Johnson, N. L., Kotz, S. and Balakrishnan, N. (1997). *Discrete Multivariate Distributions*, John Wiley & Sons, New York.

Johnson, N. L., Kotz, S. and Kemp, A. W. (1992). *Univariate Discrete Distributions*, 2nd ed., John Wiley & Sons, New York.

Jones, M. C. (2002). Student's simplest distribution, *Journal of the Royal Statistical Society, Series D*, **51**, 41–49.

Kemp, C. D. (1967). 'Stuttering-Poisson' distributions, *Journal of the Statistical and Social Enquiry Society of Ireland*, **21**, 151–157.

Kendall, D. G. (1953). Stochastic processes occuring in the theory of queues and their analysis by the method of the imbedded Markov chain, *Annals of Mathematical Statistics*, **24**, 338–354.

Klugman, S. A., Panjer, H. H. and Willmot, G. E. (1998). *Loss Models: From Data to Decisions*, John Wiley & Sons, New York.

Kolmogorov, A. N. (1933). Sulla determinazióne empìrica di una légge di distribuzióne, *Giornale di Istituto Italaliano degli Attuari*, **4**, 83–93.

Kotz, S., Balakrishnan, N. and Johnson, N. L. (2000). *Continuous Multivariate Distributions*, Vol. 1, 2nd ed., John Wiley & Sons, New York.

Kotz, S., Kozubowski, T. J. and Podgórski, K. (2001). *The Laplace Distribution and Generalizations: A Revisit with Applications to Communications, Economics, Engineering, and Finance*, Birkhäuser, Boston.

Krasner, M. and Ranulac, B. (1937). Sur une propriété des polynomes de la division du cercle, *Comptes Rendus Hebdomadaizes des Séances de l'Académie des Sciences, Paris*, **204**, 397–399.

Krysicki, W. (1999). On some new properties of the beta distribution, *Statistics & Probability Letters*, **42**, 131–137.

Lancaster, H. O. (1969). *The Chi-Squared Distribution*, John Wiley & Sons, New York.

Laplace, P. S. (1774). Mémoire sur las probabilité des causes par les évènemens, *Mémoires de Mathématique et de Physique*, **6**, 621–656.

Linnik, Yu. V. (1953). Linear forms and statistical criteria I, II, *Ukrainian Mathematical Journal*, **5**, 207–243, 247–290.

Linnik, Yu. V. (1963). Linear forms and statistical criteria I, II, in *Selected Translations in Mathematical Statistics and Probability*, Vol. 3, *American Mathematical Society*, Providence, RI, pp. 1–90.

Liouville, J. (1839). Note sur quelques intégrales définies, *Journal de Mathématiques Pures et Appliquées (Liouville's Journal)*, **4**, 225–235.

Lukacs, E. (1965). A characterization of the gamma distribution, *Annals of Mathematical Statistics*, **26**, 319–324.

Marcinkiewicz, J. (1939). Sur une propriété de la loi de Gauss, *Mathematische Zeitschrift*, **44**, 612–618.

Marsaglia, G. (1974). Extension and applications of Lukacs' characterization of the gamma distribution, in *Proceedings of the Symposium on Statistics and Related Topics*, Carleton University, Ottawa, Ontario, Canada.

Patel, J. K. and Read, C. B. (1997). *Handbook of the Normal Distribution*, 2nd ed., Marcel Dekker, New York.

Pearson, K. (1895). Contributions to the mathematical theory of evolution. II. Skew variations in homogeneous material, *Philosophical Transactions of the Royal Society of London, Series A*, **186**, 343–414.

Perks, W. F. (1932). On some experiments in the graduation of mortality statistics, *Journal of the Institute of Actuaries*, **58**, 12–57.

Pólya, G. (1930). Sur quelques points de la théorie des probabilités, *Annales de l'Institut H. Poincaré*, **1**, 117–161.

Pólya, G. (1932). Verleitung des Gauss'schen Fehlergesetzes aus einer Funktionalgleichung, *Mathematische Zeitschrift*, **18**, 185–188.

Proctor, J. W. (1987). Estimation of two generalized curves covering the Pearson system, *Proceedings of the ASA Section on Statistical Computing*, 287–292.

Raikov, D. (1937a). On a property of the polynomials of circle division, *Matematicheskii Sbornik*, **44**, 379–381.

Raikov, D. A. (1937b). On the decomposition of the Poisson law, *Doklady Akademii Nauk SSSR*, **14**, 8–11.

Raikov, D. A. (1938). On the decomposition of Gauss and Poisson Laws, *Izvestia Akademii Nauk SSSR, Serija Matematičeskie*, **2**, 91–124.

Rao, C. R. (1965). On discrete distributions arising out of methods of ascertainment, in *Classical and Contagious Discrete Distributions* (Ed., G. P. Patil), Pergamon Press, Oxford, England, pp. 320–332; see also *Sankhyā, Series A*, **27**, 311–324.

Rao, C. R. and Rubin, H. (1964). On a characterization of the Poisson distribution, *Sankhyā, Series A*, **26**, 295–298.

Sen, A. and Balakrishnan, N. (1999). Convolution of geometrics and a reliability problem, *Statistics & Probability Letters*, **43**, 421–426.

Seshadri, V. (1993). *The Inverse Gaussian Distribution: A Case Study in Exponential Families*, Oxford University Press, Oxford, England.

Seshadri, V. (1998). *The Inverse Gaussian Distribution: Statistical Theory and Applications*, Lecture Notes in Statistics No. 137, Springer-Verlag, New York.

Shepp, L. (1964). Normal functions of normal random variables, *SIAM Review*, **6**, 459–460.

Skitovitch, V. P. (1954). Linear forms of independent random variables and the normal distribution law, *Izvestia Akademii Nauk SSSR, Serija Matematičeskie*, **18**, 185–200.

Snedecor, G. W. (1934). *Calculation and Interpretation of the Analysis of Variance*, Collegiate Press, Ames, Iowa.

Srivastava, R. C. and Srivastava, A. B. L. (1970). On a characterization of the Poisson distribution, *Journal of Applied Probability*, **7**, 497–501.

Stuart, A. and Ord, J. K. (1994). *Kendall's Advanced Theory of Statistics*, Vol. I, *Distribution Theory*, 6th ed., Edward Arnold, London.

"Student" (1908). On the probable error of the mean, *Biometrika*, **6**, 1–25.

Tong, Y. L. (1990). *The Multivariate Normal Distribution*, Springer-Verlag, New York.

Tukey, J. W. (1962). The future of data analysis, *Annals of Mathematical Statistics*, **33**, 1–67.

Verhulst, P. J. (1838). Notice sur la loi que la population suit dans sons accroissement, *Cor. Mathématiques et Physique*, **10**, 113–121.

Verhulst, P. J. (1845). Recherches mathématiques sur la loi d'accroissement de la population, *Académie de Bruxelles*, **18**, 1–38.

Wimmer, G. and Altmann, G. (1999). *Thesaurus of Univaraite Discrete Probability Distributions*, STAMM Verlag, Essen, Germany.

AUTHOR INDEX

SUBJECT INDEX